Giorgio Ascoli

Biochemical and spectroscopic characterization of CP20, a protein involved in synaptic plasticity mechanism

TESI DI PERFEZIONAMENTO

SCUOLA NORMALE SUPERIORE
PISA

Tesi di perfezionamento in Chimica sostenuta il 17 dicembre 1996

COMMISSIONE GIUDICATRICE

Giuseppe Franco Bassani, Presidente
Nicoletta Berardi
Renato Colle
Pier Luigi Ipata
Dario Pini
Piero Salvadori
Antonio Tamburro

ISBN: 978-88-7642-273-7

This study was performed in the framework of a collaboration between the Centro per lo Studio delle Macromolecole Stereordinate ed Otticamente Attive del Consiglio Nazionale delle Ricerche (CNR), Dipartimento di Chimica e Chimica Industriale, Università di Pisa, Italy, and the Laboratory of Adaptive Systems, National Institute of Neurological Disorders and Stroke, National Institutes of Health (LAS, NINDS, NIH), Bethesda, MD, USA.

The purification and biochemical characterization of cp20, the fluorescence spectroscopy and the research presented in Section A of the Appendix were carried out in Bethesda. The CD and FT-IR characterization and the experiments reported in Section B of the Appendix were performed in Pisa.

The work described in Chapter 4 was carried out at the Marine Biological Laboratory (MBL), Woods Hole, MA, USA.

The purification of PKC and B-50 (Section C of the Appendix was performed at the Istituto di Scienze Farmacologiche, Università di Milano.

For my family

Acknowledgments

This Ph.D. thesis represents an important passage to me, reflecting not only scientific and professional development, but also personal growth. I want to thank all the people who were close to me in the past few years.

My mentor, Prof. Piero Salvadori, and the Scuola Normale Superiore, have together provided years of education in physical and organic chemistry, and encouraged me to go to the NIH. Dr. Dan Alkon's help went far beyond allowing me work in his lab; it was through many useful discussions with him, in addition to his papers, books, and lectures that I began to understand neuroscience.

Rebecca Goldin patiently taught me how to deal with all the aspects of the American way of life (including language). Her help in writing this thesis is only a small part of the wonderful emotional and practical support she has given me.

Dr. Jim Olds has been much more than my supervisor at the NIH. He has been a friend and a guide, always encouraging, helping, supporting: a real role model for becoming a tutor myself. Several undergraduate students assisted me in the lab; without Mark Cameron, Mary Alice Hoefler and especially Kieu Luu many results of this thesis would not have been obtained.

With Paola Pergami I shared the enthusiasm and the frustrations of dealing with a new Continent, a different life style, new relationships and... HPLC.

Several people collaborated in this work: Dr. Andrea Raffaelli, for mass spectrometry; Dr. Emilia Bramanti, for infrared spectroscopy; Dr. Carlo Bertucci and Dr. Lorenzo Di Bari, for circular dichroism; Dr. Tom Nelson, for providing many of the biochemical protocols; and Brad Zoltick, for all the computer help.

Prof. Flaminio Cattabeni, Dr. Monica Di Luca and Dr. Lucia Pastorino organized and guided my time in Milan, teaching, encouraging and always giving me their advice in the most helpful way.

I am extremely grateful to Prof. Paolo Cerletti, who stood up by my side during the transition from Chemistry to Neurosciences.

Daniel Segrè (the Gamb) was always present and close, at least electronically. I am also grateful to all my good friends: Pablo, who helped in the toughest times, Marianna, Grappy, Guido, the Kubrickclub, Claudia, Toppy, Peppe, the Italian Colony in the States, Marlene and the Great Group of Pisa...

...Thanks!

Table of Contents

List of illustrations

The experiments and the results described in this Ph. D. thesis are part of a research project which involves several disciplines and researchers. Some figures referring to experiments not performed by the candidate were included in the thesis to illustrate the context of the research. The sources of these illustrations are quoted in the legend to distinguish them from the results by the author.

CHAPTER 4

CONCLUSION   Page

APPENDIX

Giorgio A. Ascoli

List of abbreviations

ARF Adenosine diphosphate-ribosylation factor
ASW Artificial sea water
ATP Adenosine 5-triphosphate
AX Anion exchange
BCIP 5-bromo-4-chloro-3-indoyl phosphate p-toluidine salt
BSA Bovine serum albumin
CNS Central nervous system
CR Conditioned stimulus
CS Conditioned response
DTT Dithiothreitol
EDTA Ethylen-diamino tetraacetic acid
EGTA Ethylen(bis-diethylenglicol)amino tetracetic acid
GTP Guanosine 5-triphosphate
His Histidine
HPLC High-performance liquid chromatography
HSA Human serum albumin
IMAC Immobilized metal affinity chelation
Iz Imidazole
kD kilodalton
LB Louria-Bertani
LTP Long-term potentiation
mAb Monoclonal antibody
MLM Multilamellar vescicle
MW Molecular weight
NBD Methyl-nitrobenzyl-oxa-diazol-amino-dodecanoate
NBT Nitro-blue tetrazolium
OD Optical density
pAb Polyclonal antibody
PAGE Polyacrilamide gel electrophoresis
PDBU Phorbol dibutanoate
PKC Protein kinase calcium-dependent
PMSF Phenyl-methyl sulphonyl fluoride
ppAb Purified polyclonal antibody
RP Reverse phase
SDS Sodium dodecyl sulphate
TBS Tris buffer saline
TBST Tryton containing TBS
TFA Trifluoroacetic acid
t_R Retention time
US Unconditioned stimulus

Summary

Cp20 is a neuronal protein involved in the molecular mechanisms of synaptic plasticity. Cp20 was first identified in the CNS of the marine snail *Hermissenda*, which can be trained to learn an association between visual and vestibular stimuli in a Pavlovian conditioning procedure. Upon learning, cp20 is phosphorylated in a few defined neurons of Hermissenda, and inhibits the Ca^{2+}-dependent K^+ channels. This results in an increase in membrane resistance and therefore a hyperexcitability of the neural pathway. Similar biochemical mechanisms have been proposed to underlie memory in higher species, such as rabbit and rat.

Cp20 was isolated from squid optic lobes by a 5-step purification. The yield of pure product allowed the characterization of the primary structure. Cp20 was shown to have several potential post-translational modifications in addition to phosphorylation. In particular, two EF-hand Ca^{2+} binding sites were found.

The protein was cloned, fused with an oligohistidine tail, and expressed in *Escherichia coli*. Bacterial cp20 was affinity-purified and biochemically characterized as a GTP- and calcium-binding protein. The microinjection of cloned cp20 in *Hermissenda* interneurons evoked the same response observed with the microinjection of the natural protein and upon classical conditioning; the same effect was also observed in mammalian neurons.

Cp20 secondary structure was characterized by means of CD, FT-IR and sequence-matching computations. Cp20's conformation depends on the concentration of both the protein itself and free Ca^{2+} in solution. In particular, a sharp transition of the secondary structure occurs upon Ca^{2+} binding, as studied by CD, fluorescence spectroscopy and native PAGE. In the Ca^{2+}-bound form, cp20 is more globular, while the apo-protein has an elongated shape. The overall polarity does not depend on calcium. The range of Ca^{2+} concentrations at which the transition occurs is physiologically meaningful, and preliminary experiments indicate that calcium is necessary for cp20 to inhibit neuronal channels.

The activation of PKC, the enzyme responsible for cp20 phosphorylation, was also studied by an *in vivo* imaging system. PKC translocation into the cellular membrane was characterized in sea urchin eggs upon fertilization by means of laser confocal fluorescence microscopy. The translocation of PKC, as well as that of cp20, is a key passage of the activation pathway of excitable cells.

Cp20 is so far the only protein known that binds both Ca^{2+} and GTP. In light of the role that these two molecules have in neuronal signaling, and the causal relation between cp20 phosphorylation status and neuronal excitability, cp20 is a candidate for a molecular convergence point underling learning and memory.

INTRODUCTION

Pueritia quippe mea, quae iam non est, in tempore praterito est, quod iam non est; imaginem vero eius, cum eam recolo et narro, in praesenti tempore intueor, quia est adhuc in memoria mea.

Thus my boyhood, which is no longer, lies in past time, which is no longer; but as I recollect and tell my story, I am looking on its image in present time, since it is still in my memory.

(Aurelius Augustinus, "Confessiones", 399 d.c.)

1. Learning, Memory and Synaptic Plasticity

Life, in many aspects, is a remembrance of things past. Genetic memory, neuronal memory, and cultural memory shape species, individuals and cultures. Genetic memory provides organisms with innate, prefixed responses: the way *Escherichia coli*'s receptors communicate with the flagellar motors is based on memories established through evolution. The biochemical pathways that push mobile bacteria towards glucose and make them escape from its catabolites are well established in all molecular details (Hazelbauer and Harayama, 1983; McNab and Aizawa, 1984) and represent an impressive link between cause (chemistry) and effect (behavior). Some of the higher organism behaviors, however, can be modified by individual experience, and this ability to learn requires far more complicated cellular architecture: nervous systems. Neuronal learning has developed in evolution because it provides organisms with an immense adaptational potential. However, it is reductive to view brain-encoded memory like a simple evolutionary tool; some of the highest cognitive functions that characterize humans, such as abstraction, are based on associative learning, and the complexity of human behavior cannot be described in its entirety without memory. Is it possible to explain human memory on purely chemical bases, the same way bacterial chemotaxis is explained? A positive answer is suggested by the fundamental assumption of neuroscience, *that behavioral states correspond to brain states*[1]. If neuronal stimuli are the cellular correspondents of experience, and neuronal connectivity those of behavior, the ability of certain neuronal cells to modulate their connectivity in response to repeated and/or intense stimuli constitutes the cellular basis of learning and memory. This peculiar capacity is known as *synaptic plasticity*.

Neurons exchange electrical signals (information) along specialized branches, called neurites, diramating from the cellular body (the soma). Neurites form a dense communication net, which is basically an electrical circuitry of ionic currents. Each neuron has a single neurite specialized in output signals (axon) and several neurites to receive signals from other neurons (dendrites). The junctions between axons and dendrites are named synapses, and they represent the crucial points of information exchange among neurons. Synaptic activity is highly modulable and variable (hence the term synaptic plasticity); for instance, a particular pairing of signals along a determined neural pathway may activate and sensitize its synapses, thus marking that specific circuitry. This mechanism could constitute one of the cellular basis for neural memory, as extensively suggested by experimental models (e.g. Olds et al., 1990; Sunayashiki-Kusozaki et al., 1993).

---

[1]This apparently self-evident statement underlies the concept *of internal representation*, and is ardently debated by philosopher of the mind (Metzinger, 1995; Churchland and Sejnowski, 1992).

Synaptic plasticity results in neuronal changes at the anatomical as well as morphological levels (Alkon et al., 1990, and references therein). At a molecular level, synaptic plasticity is mediated by modifications of membrane receptors, cellular channels, cytostructural proteins and enzymes (for a review, see also Alkon and Nelson, 1990). In fact, it is clear that proteins play a pivotal role in the modulation of synaptic activity. Protein activity itself is controlled by several, distinct parameters that might be independently involved, such as protein compartmentalization, structure and amount. Specific proteins may be produced or over-produced as a response to neural stimuli; in particular, the induction of nucleic acids is considered one of the bases of long-term memory. RNA regulation, for instance, was reported to change with associative memory (e.g. Nelson et al., 1988 and 1990). After their synthesis by the nuclear machinery, proteins undergo a series of covalent modifications, called post-translational modifications, that influence in turn their structure and thus activity. Finally, ligand binding also modulates protein activity. It is generally accepted that nuclear induction is necessary for long term memory, while short term events are more likely regulated by post-translational modifications of proteins.

It is clear that transduction, transmission and storage of information involve a large variety of molecular steps, and the cellular physiology and chemistry that enables us to learn and recall past experience are just beginning to be understood. In the most general terms, the process of learning involves the transfer of information from our environment to networks of cells in the brain. However, given the extreme complexity of nervous systems, models are clearly needed to reduce learning and memory to simple experimental paradigms, which can be studied at the molecular level.

Several models of memory have been proposed (Alkon, 1993; Dudai, 1989; Kandel et al., 1993). With habituation, for instance, a specific neural pathway is progressively deactivated by continuous stimulation, whereas with sensitization the opposite effect is observed, and a repeated stimulus elicits a response of increasing intensity. In both examples memory acquisition consists of a mutated response by an organism to a repeatedly experienced stimulus. A widely used *in vitro* paradigm for synaptic plasticity is *long term potentiation* (LTP). When a tetanic stimulation of cortical neurons synapses is administered in specific conditions (Bliss and Collingridge, 1993; Coolly and Routtemberg, 1993), an effect similar to sensitization is observed: following the tetanus, a weaker impulse from the presynaptic neuron elicits a strong response in the postsynaptic one, as if another tetanic stimulation were applied on the synapse. LTP is not an *in vivo* model, and the experimental conditions are very far from a physiological range; therefore many neuroscientists do not consider LTP to be a paradigm for neuronal learning and memory.

Association is probably the most striking example of purely acquired information. In Pavlov's traditional paradigm (classical conditioning), distinct stimuli from the environment become associated in an animal's memory when they occur several times in temporal proximity. In Pavlovian-like experiments, a stimulus called conditioned (CS), which is usually neutral in value (i.e. causing no behavioral response), repeatedly precedes, within a limited interval of time, a second stimulus. The second stimulus, called unconditioned or US, is non-neutral, i.e. elicits a stereotypic response (called conditioned response, CR). After conditioning, the conditioned response is also elicited by the previously neutral conditioned stimulus[2] (Pavlov, 1910). An accurate identification of memory-specific molecular steps requires careful comparison of learning experience to identical sensory and motor experience in the absence of learning. In other words, if we define memory as information storage, we should find an experimental paradigm suitable to isolate the informational content component. Associative learning is particularly amenable to this kind of design. There is a natural control paradigm, mainly the random (unpaired) sequence of both stimuli presented to the animal. Any biophysical, behavioral, biochemical or structural changes observed after training with temporally paired stimuli and not with the control unpaired ones can then be considered memory-specific. In addition, associative memory is not only responsible for many acquired reflexes in animal behavior, but also for some of the highest cognitive functions, such as language; when a parent presents a child with object names, the US consists of the site of the object, whereas the sound of the name represents the CS.

2. The *Hermissenda* model: from behavior to proteins

Pavlov's original description of classical conditioning regarded dogs, and today it is widely accepted that virtually every mammal shows associative abilities. While a huge research effort was directed in the past decades towards higher species (monkeys and even humans) or easily controllable habituation and sensitization paradigms (e.g. Kandel, 1976), an alternative strategy is to seek the simplest species able to associate independent stimuli. It would be particularly advantageous to find such a species amongst invertebrates. Their nervous systems and sensory structures are in fact especially suitable for an analysis of the biochemical mechanisms involved in behavioral modification because they typically contain a small number of neurons, some of which have large, identifiable somata. Studies with the nudibranch *Hermissenda crassicornis* (fig. 1), a marine snail originally found in the Pacific Ocean, demonstrated a neural connection between its only two sensorial pathways, tactile and visual. A short-term non-

---

[2] Logically, the CS is a conditionable, rather than conditioned, stimulus. In fact CS becomes conditioned during the experiment.

associative change in phototactic response (the animal instinct to move towards light) was originally observed after training with light and rotation.

Figure 1. *Hermissenda Crassicornis (modified from the cover of Science News, vol. 139, N. 21, May 25, 1991). The animal is about 2 cm long.*

Hermissenda interpreted rotation as an aversive stimulus (contracting as if it were responding to water turbulence) and delayed its movement towards illuminated regions (Alkon, 1974). A paradigm of associative learning was subsequently produced by temporal association of the two stimuli: after specific training procedures, Hermissenda could learn that a light flash was followed by violent turbulence, and started contracting in reaction to light (Crow and Alkon, 1978). The emergence of this new behavioral response showed all the features of classical conditioning: retention of the learned response for up to 14 days after the end of training (Crow and Alkon, 1980), extinction, and savings on retraining (Matzel et al., 1992).

The entire Hermissenda nervous system was mapped by means of diffusible dyes and intracellular recording. Among the several hundred cells, the type B photoreceptor neuron in the eye was found to be the only storage site of the acquired memory (Crow and Alkon, 1980). In particular, the weak connection between the visual and neuromuscular pathway, which elicits no defensive response to light in non-trained animals, became heavily activated in trained

Hermissendas[3] (fig. 2). Electrophysiological studies demonstrated that a reduction of the two K^+ currents I_A and $I_{Ca2+-K+}$ in type B photoreceptor is precisely responsible for this activation (Alkon et al., 1982; Alkon, 1984); the modification of potassium channel activity (which is normally used to repolarise the neuron after firing, i.e. to switch it off) has been shown to leave the neuron in an active state for weeks. Such an effect is necessary for memory retention and sufficient to elicit the behavioral effects of conditioning (Farley et al., 1983; Alkon et al., 1985).

Once a specific cellular response was characterized as the neural correspondent of the learned behavior, a molecular approach was undertaken. Evidence was found that kinase activity mediated the electrophysiological changes underlying learning and memory, i.e. the phosphorylation of specific proteins was a natural candidate for playing the role of marker and functional modulator of cellular activity (Alkon, 1979). This metabolic pathway was subsequently identified as that of the enzyme PKC (Farley and Auerbach, 1986), and microinjection of PKC in type B photoreceptors evoked the same reduction of K^+ channels as classical conditioning (Acosta-Urquidi et al., 1984; Alkon et al., 1983 and 1988).

The reduction of calcium-dependent potassium currents and the activation of PKC are not specific phenomena of the Hermissenda's nervous system; on the contrary they represent a very general "trace" of associative memory in many if not all organisms able to show Pavlovian conditioning. Hyperpolarization and prolonged activation of neurons mediated by inhibition of potassium channels were also observed and fully characterized in rabbit hyppocampus and cerebellum after associative learning (for a review, Schreus, 1989), and these same mechanisms are defective in human cells from individuals with memory impairment, such as Alzheimer Disease patients (Etcheberrigaray et al., 1994). PKC is involved in virtually all memory experimental models, from long-term potentiation (Malenka et al., 1987; Lovinger and Routtemberg, 1988) to habituation (Sacktor and Schwartz, 1990). In the rabbit hyppocampus, PKC activation following associative conditioning modifies ion channels, causing a reduction of the calcium-dependent potassium currents (Alkon and Rasmussen, 1988; Kazemerak, 1992; Lester and Alkon, 1991).

---

[3] Interstingly, a neural network (Dystal) was developed based on the Hermissenda nervous system. For references, see Alkon, 1995.

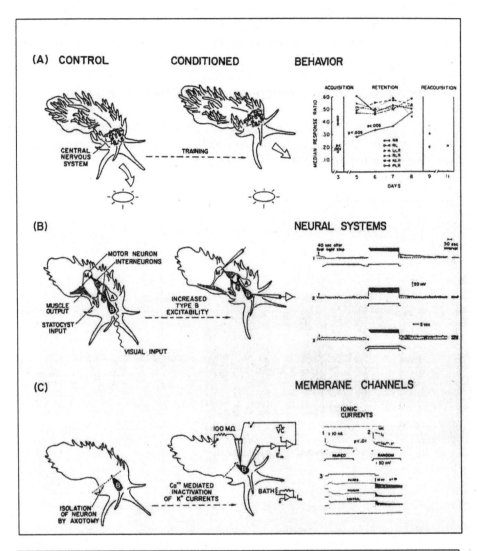

Figure 2. *Associative conditioning in Hermissenda. (A) Behavioral studies: the animal has a natural positive phototactive response, which is transformed upon training with an aversive stimulus. Learning and memory follow Pavlovian classical parameters of acquisition, retention and reacquisition. (B) Neural connectivity: Hermissenda nervous system is mapped and each cell is electrophysiologically characterized. Only type B interneurons are changed by associative conditioning. (C) Intracellular recording: the inactivation of calcium-dependent potassium channels is proved to be responsible for the increased excitability of type B interneurons (adapted from Alkon, 1993).*

The hypothesis that PKC phosphorylates a substrate, which in turn causes the reduction of $I_{Ca2+-K+}$, represents a plausible and widely general molecular mechanism for associative learning, although many details remain to be explained.

3. Physiological and biochemical properties of cp20

The interest in PKC substrates in Hermissenda began in 1981 with the observation that a specific 20 kD-protein in the snail's eye was phosphorylated upon associative learning (fig. 3; Neary et al., 1981).

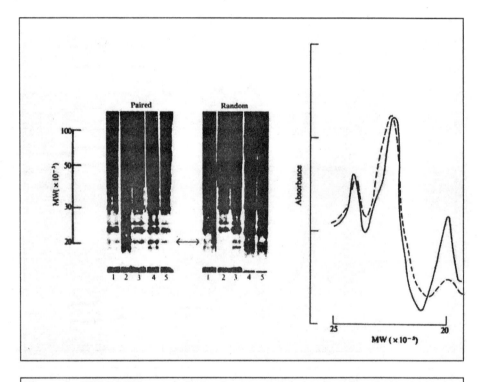

Figure 3. *SDS-PAGE comparing endogenous protein phosphorylation in eyes of Hermissenda presented with paired or random light and rotation. Each lane of the gel represents an eye sample from one animal. High molecular weight extensively labeled phosphoprotein bands were not significantly different between the two groups. As shown in the densitometric scan, there is a significant difference at 20 kD. Full line: paired; dotted line: random. Absorbance is in arbitrary unit. Adapted from Neary et al. (1981).*

The only useful element for the isolation of this molecule (from a tissue, the Hermissenda eye, visible only by microscope), was that it is phosphorylated upon associative conditioning. The protein was identified after nine years by analytical HPLC (fig. 4). The phosphorylation adds a negative charge to the protein, and therefore affects its affinity for the anion exchange resin. The protein extract of trained Hermissenda's eye shows a different chromatographic pattern from that of the naive animal's one. One of the <u>c</u>hanged <u>p</u>eaks contained a protein of <u>20</u> kD (*cp20*), which was identified as a GTP-binding protein (Nelson et al., 1990).

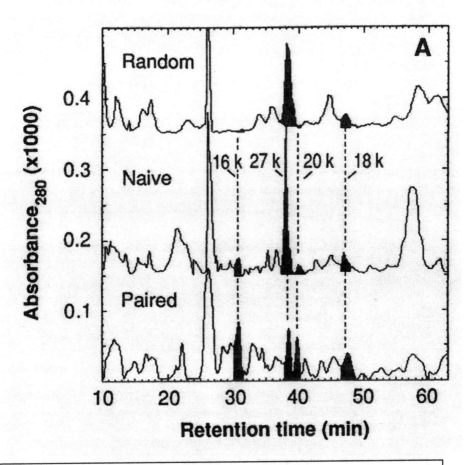

Figure 4. *Typical analytical AX-300 HPLC tracings of proteins from eye of a Hermissenda subjected to light and rotation (upper), a naive animal (middle), or an animal subjected to light-rotation pairing (lower). Zero is set at 0.35 for the random tracing, 0.15 for naive, and 0.0 for paired. Unmarked peaks were variable among animals. 85% of the injected proteins eluted in the non-retained fraction between 0 and 10 minutes. From Nelson et al., 1990.*

When the cp20 peak was microinjected in the B photoreceptor of Hermissenda, it caused the same inhibition of K^+ channels found after associative conditioning.

Although the discovery of cp20 represents one of the first examples of direct correlation between a molecular entity and a behavioral effect in the field of associative learning and memory, the amount of isolated protein was barely detectable. A Hermissenda eye contains only a few micrograms of protein, and subnanogram quantities of cp20. Among relatively larger animals genetically related to Hermissenda, the squid shows a very similar 2D-gel protein pattern. A crude extract of squid optic lobe was chromatographed on the analytical anion exchange HPLC (fig. 5), and the retention time of all the peaks correlated linearly with the conditioned Hermissenda chromatogram (Nelson et al., 1994). The squid peak corresponding to Hermissenda's cp20 contained a 20 kD protein that affected potassium current when microinjected in Hermissenda interneurons.

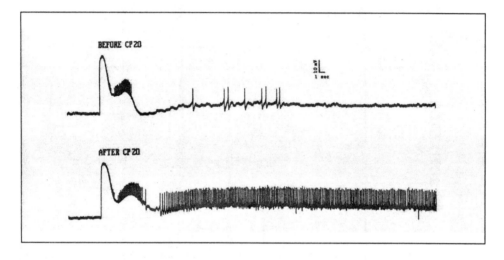

Figure 5. *Intracellular recording of the effect of cp20 on the light response of Hermissenda type B photoreceptor interneurons. Iontophoretically microinjected cp20 caused a marked increase in excitability, which was traced to inhibition of outward K+ currents, whereas the Na+ currents (the initial broad peak) and resting potential were unaffected (Nelson and Alkon, 1990).*

With a partial purification protocol developed from the same HPLC system, enough protein was isolated from squid optic lobes to allow a preliminary characterization (Nelson et al. 1994). The molecular weight measured by sodium dodecyl-sulfate polyacrilamide gel electrophoresis (SDS-PAGE) was confirmed by size exclusion HPLC, and the isoelectric point was determined to be 5.1 by isoelectrofocusing and chromatofocusing. Squid cp20 showed GTPase activity, and its iontophoretic injection in Hermissenda type B photoreceptor and rabbit hippocampus neurons caused the potassium current reduction previously correlated with associative learning (fig. 6; Olds and Alkon, 1991).

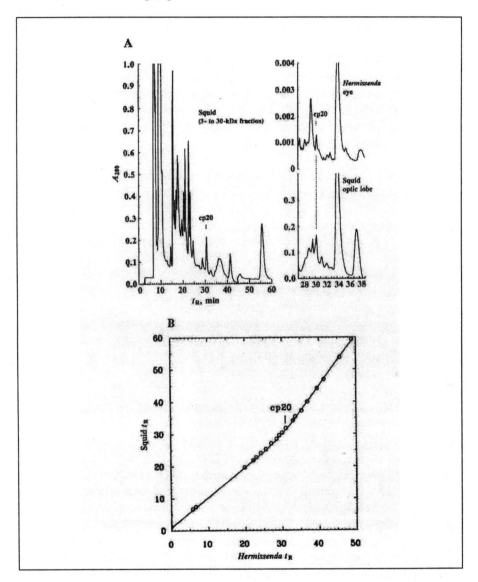

Figure 6. *(A) Semi-preparative AX-300 HPLC tracings of proteins from Hermissenda eye, squid optic lobe and squid 3- to 30-kD fraction. (B) Correlation curve of t_R values from squid optic lobe proteins and trained Hermissenda eye reference chromatogram (from Nelson et al., 1994).*

Cp20 also induced morphological changes of the synaptic terminals indistinguishable from those observed with conditioning (Nelson and Alkon,

1991). Furthermore, purified cp20 was phosphorylated *in vitro* by PKC, and was proved to translocate into the cellular membrane upon phosphorylation. Finally, cp20 inhibited axonal transport in a similar way to other low molecular weight neuronal G-proteins, possibly depending on its phosphorylation state (Nelson and Alkon, 1995).

Other 20 kD phospho-proteins in the same family as cp20 were successively identified in mammals such as rabbit and human as well as in invertebrates such as sea urchin (Nelson et al., 1994, 1995). Cp20 was shown to be reduced both in rabbits after associative conditioning (Nelson et al., 1991) and in humans affected by Alzheimer disease (Kim et al., 1995).

Cp20 clearly represents a unique and robust candidate for a chemical effector of the physiological changes underlying learning and memory. The effects of cp20 on neuronal cells are similar to those observed after associative training, and its phosphorylation state is the only molecular parameter causally related to classical conditioning. A partial action mechanism could be exemplified by the following model (Alkon, 1995): upon conditioning, PKC is activated and phosphorylates cp20. This causes cp20 translocation to the membrane, where it inhibits calcium-dependent potassium channels, reducing the related currents and prolonging the neuron's activity. The plausibility of such a model is obviously to be demonstrated, since the molecular details of cp20 interactions are not yet known. In particular, all the cellular elements interplaying with cp20 should be individuated, and their structural parameters characterized.

4. Spectrometric techniques and protein structures

Protein structure is classically divided into several levels. The primary structure is the aminoacidic chain sequence, translated from RNA and ultimately determined by the DNA genetic code. Primary structure determines the protein identity but not necessarily its activity. After the synthesis, in fact, proteins undergo a series of reactions (usually carried out in the Golgi apparatus), such as phosphorylation, acylation and removal of peptidic portions. The protein is then brought to its proper cellular or extra-cellular compartment, and is further modified by specific enzymes. These reactions are called post-translational modifications in that they are not directly coded by the species genome. Post-translational modifications contribute to the determination of protein's three-dimensional structure and its properties. In neural cells, post-translational modifications are commonly used as switches to mark the protein; typical examples are the phosphorylation and the acylation which determine protein location and activation (see e.g. Di Luca, 1992; Alkon et al., 1988).

The secondary structure is the local disposition of parts of the proteic chain in the space. Several stable secondary structures are known, such as α-helix and β-sheet. The overall conformation of the protein, based on its local secondary structures, is called tertiary structure. The properties of a protein can be altered if it switches from one tertiary structure to another; inducing a change in the tertiary structure is one mechanism of protein activity regulation. Binding areas and interaction sites, the active portions of the protein, are formed according to the protein's three-dimensional structure. It is worth underlining that protein binding of other proteins or small ligands both influences and is influenced by the protein structure. Enzymatic catalysis, pharmacological action and metabolite transport, are all examples of protein activities mediated by binding properties.

The modulation of protein properties is deeply related to its structure at several levels, and it is just the complexity of such a structure that allows the accomplishment of a large variety of biological functions. One of the most important aims in neuroscience is therefore the understanding of the relationship between structure and activity of the proteins responsible for the phenomenon of synaptic plasticity. Classical techniques generally used in neurobiology, such as intracellular recording, microinjection, immunostaining and western blotting, can help to recognize which proteins are involved in specific *in vivo* cellular effects, such as morphological changes or ionic channel regulations. Protein structure analysis in contrast must be carried out *in vitro*, on purified or partially purified proteins.

The study of protein structure is usually accomplished by chemical or chemical-physical techniques such as Edmand degradation for primary structure (although molecular biological techniques of DNA cloning and sequencing are routinely used in parallel), X-ray crystallography for three-dimensional structure, and, with increasing frequency, spectrometric techniques. Spectrometries present several advantages such as velocity and conceptual simplicity, although the technology to analyze large biopolymers was developed only recently. In particular, post-translational modifications, secondary structures and binding properties can now be studied, among others, by mass spectrometry, fluorescence, Fourier-transform infrared spectroscopy and circular dichroism.

Until recently, a major problem in the study of post-translational modifications consisted in the impossibility to detect them by Edmand degradation under normal conditions; the covalent modifications of the protein are generally removed in the process of fragmentation. Molecular biology techniques are clearly not suitable to study such modifications, since they are not coded in the DNA. In the past few years, fast atom bombardment mass spectrometry (FAB-MS) proved useful in studying post-translational modification of neural protein fragments (for a recent application, see Di Luca et al., 1992), but the main limitation lay in the low molecular weight limit of the peptides that could be analyzed. With the development of ion spray mass spectrometry

(IS-MS) and tandem mass spectrometry (MS/MS) these limitations were bypassed and the determination of the exact molecular weight of entire proteins became possible (Chait and Kent, 1992). If the protein is post-translationally modified, the molecular weight shifts, and from such a change it is possible to determine the kind of modification. Coupling these techniques with HPLC allows the direct analysis of partially purified mixtures, quite common in biological matrices (Nuwaysir and Stultz, 1993). More recently, even protein binding interactions were detected by IS-MS (Huang et al., 1993).

Fluorescence was historically the first spectrometric technique used to gain insight into protein structure (Wang et al., 1992; Linse et al., 1987 and references therein). Although fluorescence excitation spectra are sometimes recorded to optimize the experimental conditions, protein structural analysis is usually carried out by two types of emission spectroscopy: intrinsic and extrinsic. In intrinsic fluorescence, the protein is excited at a suitable wavelength for the aromatic residues tryptophan, tyrosine and phenylalanine (usually around 280 nm), and the emission is recorded. The indolic group of tryptophan yields a much stronger signal than the aromatic aminoacids, but tyrosine and phenylalanine are usually more abundant in the aminoacidic sequence. Anyway, it is possible to distinguish the contribution to the emission of the various aminoacids by changing the excitation wavelength (since they have different absorption maxima). The intensity of the signal depends on the exposure of the excited aminoacids; therefore, if there is a conformational change in the protein, the fluorescence spectrum can detect it. Furthermore, if only few residues of tryptophan are present in the studied protein, the fluorescence spectrum change can be correlated to a structural transition in proximity of the indolic aminoacids. This same idea is applied in extrinsic fluorescence emission spectroscopy. In this case, a fluorescence ligand is bound (covalently or by complexation) to the protein, and it constitutes a probe at transparent wavelength. Such a molecular antenna can be used in microscopy imaging as well: if a fluorescent molecule is condensed with a specific ligand for a particular protein, its injection (or diffusion) in the cell sensitively and selectively marks that protein. Although fluorescence spectroscopy usually yields only qualitative information, it is widely diffuse for its simplicity. In addition, because it requires the use of diluted solutions, samples recovered from other spectroscopic analyses can be used, with remarkable material conservation.

Infrared (IR) spectroscopy was one of the earliest experimental methods that proved to be useful in evaluating the secondary structure of polypetides and proteins (Parker, 1983). Several problems connected with the vibrational analysis of proteins, such as the low sensitivity of IR instruments and interfering absorption of the solvent, were largely overcome by the employment of Fourier transform IR (FT-IR) spectrophotometers, which made it possible to acquire high-quality spectra from amounts of proteins in the order of 10 μg, in a variety of environments (Hadden et al., 1995; Costantino et al., 1995). The amide group of proteins presents characteristic amide vibrational modes that are sensitive to conformation. Amide I (1700-1600 cm^{-1}) is primarily due to the C=O stretching vibration, while amide II (1600-1480 cm^{-1}) and amide III (1350-1190 cm^{-1}) result from the coupling of different stretching modes. The difference in amide bond geometric orientation of α-helix, β-sheet, turn and random coil structures allow for differences in vibrational frequencies and distinct IR signals. The assignment of the amide component bands to different types of secondary structures is a critical step in the interpretation of IR spectra of proteins. This assignment is guided by theoretical calculations and by spectra-structure correlations established experimentally for models of polypeptides and proteins whose three-dimensional structures are known. In particular, the assignment of amide I components is widely used because of the intense protein signal. The amide II can only be used for qualitative confirmations, due to the simultaneous absorption by aminoacid side chains in the region. Amide III analysis can also be used as a complementary method to amide I analysis in protein structural studies (Fu et al., 1994). A recent method based on the conjugate gradient minimization algorithm, and the application of the second derivative operator, determines the central band frequency, bandwidth, and amplitude of the different spectral components (Bramanti and Benedetti, 1996). Thus such a method makes it possible to calculate quantitatively the contribution of each component to the overall spectral absorption. By knowing the assignment of these components to the different secondary structures, it is possible to evaluate the conformations of the protein domains. FT-IR has the additional advantage of allowing the analysis of spectra taken either in solution or in solid state, over a wide range of concentrations. It is therefore a suitable technique to study aggregation phenomena.

Protein secondary structures can also be studied by circular dichroism, as described in more detail in the following paragraph.

5. Circular dichroism and secondary structure

Circular dichroism spectroscopy (CD) can be used to analyze protein secondary structures in solution, i.e. under conditions closer to its physiological environment with respect, for instance to those typical of X-ray studies. On the other hand, CD yields less detailed information on the protein's conformation than crystallography. CD analysis of protein secondary structure requires specific technical devices to carry out measurements at a relatively high energy of polarized light beam. The peptidic chromophore absorption, in fact, results in different short-wavelength CD bands depending on the conformational content. Data analysis processing, standard reference spectra and experimental details have been extensively reviewed (Johnson, 1990; Manning 1989; Woody, 1985).

In the most simple approach, an experimental far-UV (250-175 nm) CD spectrum is deconvoluted into a linear combination of basis spectra, each representative of a secondary structure type. The relative contribution of each basis spectrum to the overall CD curve is proportional to the amount of that structural feature in the protein. Basis spectra are obtained by polypeptides known to be in a single ordered structure. This method presents some problems that make it partially unreliable. First of all, the CD spectrum of an α-helical polypeptide represents an extremely long peptide, while only short portions of α-helix are present in proteins (the same is true for the other structures). Second, some contributes to protein CD, such as lateral chromophoric chains, cystinic bridges and secondary structure distortions, are not represented by polypeptides. Third, only a few secondary structures can be obtained in large excess in polypeptides. These problems are resolved by obtaining basis spectra from real proteins having a known secondary structure composition (as determined by X-ray christallography). For this purpose, one of many variable curve-fitting routines may be employed (Hennessey and Johnson, 1981; Woody, 1985). A typical example of this approach is given by the *singular value decomposition* (SVD) of Johnson and coworkers (Johnson, 1992, and references therein). The CD spectra of 33 proteins form a matrix **R** of 33 rows and 83 columns (the number of data points obtained by taking the molar ellipticity at every nm over the range 260-178 for each of 33 proteins). Secondary structure content form a second matrix, **F**, which is 33 X 5 (assuming there are five different secondary structures which can be resolved). Extension of the data into the vacuum UV ($\lambda < 180$ nm) seems to provide greater information content and improved accuracy. Thus, the matrix **X** is sought, defined by:

$$F = X R$$

The SVD framework proceeds via decomposition of the matrix **R** into the product of three different matrices (for a complete mathematical treatment, see e.g. Manning, 1989), which contain orthogonal basis vectors (the eigenvectors of the product of **R** and its transpose). The transformation matrix, **X**, can be then used to obtain the amounts of any of the five secondary structure types, namely α-helix, parallel β-sheet, antiparallel β-sheet, turns and random coil, or unordered, conformations.

However sophisticated the mathematical treatment is, secondary structure analysis by CD involves numerous assumptions, many of which are independent of the deconvolution method. While the accuracy of various basis sets (i.e. reference spectra) can be examined systematically (Johnson, 1990), the following assumptions should be considered. First, the effect of tertiary structure is negligible. This means that individual secondary structural elements do not interact and their contribution to the overall CD spectrum is additive. Second, only the amide cromophores are responsible for the specific far UV CD spectrum, i.e. contributions from side chain cromophores are assumed to be averaged by the basis proteins. Third, the geometric variability of secondary structures is assumed to be negligible (a single CD curve is sufficient to describe each type of secondary structure). An extensive discussion of these points can be found in critical reviews (e.g. Manning, 1989).

At lower energy, protein CD spectra show generally no absorption band, and this property can be used to study stereospecific ligand binding: since protein binding areas or sites are usually dissymetric, they can bind prochiral ligands in a preferred optically active conformation, and/or discriminate between enantiomers of chiral ligands. In these instances, if the ligand absorbs at a long wavelength, a new CD band could appear (in case of a prochiral molecule) or the usual one for a chiral compound could be shifted due to a polarity change of the medium. By subtracting the spectra of the components from the spectrum of the complex, a difference signal is thus obtained which accounts for stereospecific interactions (see for example Ascoli et al., 1995a; Bertucci et al., 1990).

6. Aim of the project and organization of the thesis

The present research project aims to contribute to the study of the correlation between structure and biochemical activity of proteins involved in the cellular mechanisms of synaptic plasticity. In particular, the protein cp20, which represents a crucial pathway for many physiological effects of learning and memory, is studied. The characterization of cp20 primary and secondary structures is the main accomplishment of the thesis, and required a relatively large amount (0.1 mg) of protein.

A new purification protocol from squid optic lobes was implemented, which allowed the recovery of a few micrograms of pure protein (over a ten-fold improvement from the previous protocol). This medium-scale purification, described in Chapter 1, required one high-speed centrifugation, two ultrafiltrations, and three chromatographic steps. The purified squid protein was used to obtain the aminoacidic sequence of some proteolytic fragments. A clone was built from such sequences, and the entire cp20 sequence was finally found in a squid genomic library. In order to produce a larger quantity of cloned protein, an E. coli strand was infected with a lactose-activated gene for cp20, fused with an oligo-histidine tail (to allow an easier purification by metal-chelated affinity).

The primary structure of cp20 is described and discussed at the beginning of Chapter 2. Several consensus sequences for potential post-translational modifications (including PKC-phosphorylation) are noted, and in particular the presence of two EF-hand calcium binding domains is suggested. Chapter 2 describes the purification of cloned cp20 (also called calexcitin) from E. coli, and a preliminary biochemical characterization, which confirms the identity with the squid protein. Calcium binding activity was also detected.

The complete spectroscopic characterization of cp20 secondary structure is reported in Chapter 3. FT-IR, theoretical calculations and CD yielded results in excellent agreement with each other. In particular, the CD analysis detected a conformational switch of the protein upon calcium binding. Such a transition (constituted by an increase of α-helix content and a reduction of β-sheet components) was also confirmed by intrinsic emission fluorescence spectroscopy and non-denaturating gel electrophoresis. In addition, chromatographic experiments suggested that the polarity of the protein is not influenced by calcium. The last paragraph of Chapter 3 describes two chemical modifications carried out on cp20, i.e. the histidine-tail removal and the phosphorylation by PKC. While the first reaction affected the molecular weight, but not the structural properties of cp20, preliminary results indicate that the PKC phosphorylation seems to stabilize the calcium-binding structure.

Chapter 4 describes a relatively independent research line, the imaging of the compartmentalization of the enzyme PKC in living cells. PKC is activated by translocating into the membrane, and the PKC phosphorylation triggers the translocation of cp20 itself into the membrane, where it could act on the calcium-dependent potassium channels. Therefore an *in vivo* imaging system for PKC based on fluorescence confocal microscopy is required to completely elucidate the action mechanism of cp20.

All chapters are divided (after a brief introduction and summary) into a *Materials and Methods* section and a *Results and Discussion* section. In general, when the same technical procedures have already been used for experiments described earlier in the thesis, the *Materials and Methods* section refers to the appropriate previous chapters. The chapter of Conclusions summarizes the most original and suggestive findings in a speculative but simple model of cp20 activation, and outlines future research directions. The bibliography is ordered alphabetically in the last section.

An appendix divided in several paragraphs is included in the present thesis. It describes research lines not directly related to the main project, but also carried out during the Ph.D. research period. Appendix paragraphs do not follow any particular internal or external order. Appendix A presents a mathematical model to describe dendritic spines in a vector space. The model introduces a new coordinate system that allows, for instance, an easy description of the delivery of new proteins or genes from the soma to the dendrite. The problem of how dendritic spines are targeted by the soma is still open in Neuroscience, but a theory has been proposed that involves cp20 (the *zip-coding hypothesis*). Appendix B describes the preparatory work carried out on a commercial protein, human serum albumin (HSA), to test the potential of some techniques utilized in the main project; several post-translational modifications were induced on HSA, and circular dichroism was used both to test the secondary structure and to study the stereospecific binding properties (i.e. the biochemical activity) of the modified proteins. Appendix C describes the preliminary results obtained in the analysis of another protein involved in synaptic plasticity mechanisms, B50. Like cp20, B50 is phosphorylated by PKC, and presents other post-translational modifications as well. The attempt to implement an analytical method based on the on-line chromatography-mass spectrometry separation of single-brain extract is also described.

CHAPTER 1

Medium-scale purification of squid cp20

After the original observations of cp20 phosphorylation in Hermissenda eye with associative learning, it was necessary to use a different source to obtain a larger quantity of cp20 for a more complete characterization. Cp20 had been found in many other species (sea urchin, octopus, shrimp, rabbit, rat, mouse), and among them, squid showed a genetic structure similar to that of Hermissenda, with a protein profile of the optic lobe almost identical to that of Hermissenda eye on a 2-D gel. In addition, at least 50% of cp20 found in squid optic lobes is water soluble, similar to cp20 in Hermissenda but in contrast to that of rabbit, for instance, whose cp20 is found almost entirely in the membrane fraction.. The possibility of extraction without detergents is clearly favorable since detergents may interfere with other steps of the purification, as well as with the spectrometric analysis. The original purification protocol for squid cp20 was based on a anion exchange (AX) HPLC fractionation after 3/30 kD double molecular weight cut-off of the water-soluble proteins, extracted in a dithiothreitol (DTT)-containing buffer (Nelson et al., 1994). DTT was added to prevent cp20 dimerization, thus increasing the efficiency of the following ultrafiltration passage. A remarkable loss was nonetheless observed in this step, probably due to a non-specific absorption onto the filtration membrane. An immunoaffinity purification method was then attempted.

A short proteolytic fragment of partially purified cp20 was sequenced, and a rabbit polyclonal antibody (pAb) was raised against the synthetic peptide. Unfortunately, the partial sequence was insufficient to succeed in a study of the primary structure by either molecular biology techniques or computerized genetic library screening. A mouse monoclonal antibody (mAb) was also raised against the gel-extracted protein. The antibodies cross-reacted with other low-molecular weight G-proteins of the ARF super-family, and the partial sequence actually showed 50% homology between cp20 and some of these proteins. The pAb also recognized a form of rabbit cp20, while both antibodies detected Hermissenda and human cp20, as verified by HPLC dot-blot and western blot. The antibodies were unstable though, since their immunoactivity remarkably lost specificity over 10 months. A multiple affinity chromatography purification of cp20 was nonetheless tried, as described in this chapter. The pAb was affinity purified with the antigenic synthetic peptide, and immobilized on a stationary phase. The column, however, failed to retain cp20 from squid optic lobe supernatant. By means of a similar technique, the mAb was also immobilized on a stationary phase. Both squid optic lobe supernatant and rabbit brain urea extract[1] were loaded on the column. After desalting, the retained fractions were passed through a GTP-agarose stationary phase. None of the resulting peaks contained detectable amounts of cp20.

Finally, a new strategy was undertaken, and a semipreparative protocol established. After homogenization, centrifugation and ultrafiltration, a large amount of tissue (over

---

[1] The mAb antibody was raised against denatured cp20, so the antigenic epitope could lye on an internal pocked of the protein. This justifies the use of urea, to facilitate a complete unfolding of the membrane-extracted proteins.

400 squid optic lobes) was loaded on a preparative low-pressure weak anion-exchange column. The cp20-containing fraction was recovered, diluted and loaded by continuous flow on a Cibacrom Blue medium-pressure column. The AX-HPLC step from the previous protocol (Nelson et al., 1994) was left as a final purification passage, because it allowed the recognition of the cp20-containing peak by computer-assisted peak matching, without the use of antibodies. By such a protocol, several micrograms of cp20 could be isolated, and the purity could be increased up to more than 95% by reinjection in AX-300 HPLC.

A proteolytic digestion of purified cp20 was separated by capillary electrophoresis and the fragments sequenced by Edmand degradation. Several nucleotidic primers were synthesized corresponding to the oligopeptides, and a clone was produced. The entire cp20 gene was finally individuated by screening the squid genetic libraries.

Materials and Methods

Optic lobes from fresh squid (*Loligo pealei*; Calamari, Inc., Woods Hole, MA, USA) were dissected on dry ice under sterile conditions, frozen in liquid nitrogen, and stored at -80°C. All chemicals were reagent grade and of the highest commercially available purity. Protein concentration was determined with the Lowry assay (Bio-Rad, Richmond, CA, USA), according to the manufacturer's protocol, or with the bincichonic acid test (Smith et al., 1985) using bovine serum albumin (BSA) as a standard.

All the high-pressure chromatographic steps were performed on a Becmkan System Gold equipped with a 266 detector (Becmkan, Fullerton, CA, USA) and controlled by an IBM PS/2 56SX computer. The absorbance was monitored at 280 nm. All solutions used in HPLC were filtered through a 0.2-mm nylon filter (Nalge Company, Rochester, NY, USA) and degassed immediately prior to use. The low- and medium-pressure chromatographic steps were performed either on the same Becmkan system or on a Pharmacia SX10 peristaltic pump (Pharmacia, Uppsala, Sweden). Fractions were collected and the absorbance at 280 nm measured on a Shimadzu desk-top spectrophotometer (Shimadzu Co., Tokyo, Japan). All columns were regenerated after each preparation following the manufacturer's indications. All aqueous buffers and HPLC solutions were based on ultrapure deionized water (Aquapore system, Las, NIH, Bethesda, MD, USA) containing 0.03% NaN_3 as a bactericide.

Sodium dodecyl sulfate polyacrilamide gel electrophoresis (SDS-PAGE) was performed on 4-20% gradient precasted gels (Novex, S. Diego, CA, USA), following the manufacturer's protocol. Briefly, samples containing maximum 30 µg of proteins (at an ionic strength of less than 0.4M) were mixed with 3x sample buffer (3.25 ml of 0.5 M trisCl, pH 6.8, 3 ml of glycerol, 6 ml of 10% SDS, 1.25 ml of β-mercaptoethanol, 0.75 ml of 0.005% bromo-phenol blue), heated at 100°C for 4 minutes, centrifuged on a bench-top centrifuge (14000 rpm, Becmkan) and loaded in commercial Novex running buffer. Low-range molecular weight markers were supplied through Boehringer Mannheim (Indianapolis, IN, USA). The gels were silver stained (Intense BL, Amersham, NY, USA), scanned and analyzed with an MCID/M2 imaging system (Imaging Research, St. Catherines, ON, USA). Known amounts of BSA, myoglobin, cytocrom C and histone were separated by SDS-PAGE; after silver staining, a standard curve was generated for correlating optical intensity of samples to their actual protein content. Fractions collected from different steps of the purification were quantitated by this method, only using values in the linear range of the curve.

Dot-blot analyses were performed in an 80-sample commercial apparatus (Bio-Rad, Richmond, CA, USA); 400 μl of sample were loaded (twice in some experiments) under light vacuum. When all the solutions was through, the membrane was soaked in water for 10 minutes, air dried and stored at 4°C overnight. For primary antibody detection the nitrocellulose sheet was simply incubated with secondary antibody, washed and stained. For cp20 detection, the western blot procedure described in the Materials and Methods of chapter 2 was used.

All ultrafiltration membranes (Centricon C10 and C50, Centriprep C10 and C50; Amicon, Beverly MA, USA) were pretreated with BSA. Use of untreated membranes led to complete loss of cp20 due to non-specific absorption. A solution of 1% BSA was ultrafiltered, the membrane washed 20 times with water, then distilled water was ultrafiltered, and the membrane was washed again 20 times, and stored in water at 4°C. All ultrafiltration and concentration steps were performed at 15°C, at the speed and angle indicated by the manufacturer.

The small-scale purification of cp20 followed a procedure modified from a previously used protocol (Nelson et al., 1994). Briefly, 30 optic lobes were added to 20 ml of sterile loading buffer (50 mM trisCl, pH 7.4/1mM EDTA/1 mM EGTA) containing 1 x protease inhibitor cocktail (20 mg/l pepstatin/20 mg/l leupeptin/10 mg/l aprotinin/0.5 mM phenylmethylsulfonyl fluoride, PMSF; the cocktail was centrifuged at 14000 x g - Sorvall SS34 rotor - to remove insoluble matter). 50 mM NaF and 200 mM DTT were added to prevent phosphatase activity and dimerization, respectively. The optic lobes were homogenized at 0°C by high-power sonication (TJN, Las, Bethesda, MD, USA). The crude extract was centrifuged (100000 x g, 90 minutes, 4°C) on a high-speed table-top ultracentrifuge (Becmkan T150 rotor). The supernatant was ultrafiltered through Centricon, and the low molecular weight fraction was then concentrated down to 100 μl on Centricon C10. The retained fraction was injected with a 100 μl loop onto a semi-preparative AX-300 anion exchange HPLC column (1 x 25 cm, 30 nm gel particle diameter; Thomson, Springfield, VA, USA). The column was eluted at 2 ml/min and 8°C with a linear gradient of 0-0.6 M potassium acetate buffer (pH 7.4) for 20 minutes followed by 0.6 M buffer for 30 minutes and 1 M buffer for 10 minutes. Each chromatogram was statistically analyzed by creating a correlation curve with the retention time (t_R) of each peak plotted against the t_R of all peaks in a reference chromatogram, a chromatogram of proteins from five eyes dissected from a group of Hermissendas conditioned in a previously described experiment (Nelson et al., 1990). A candidate cp20 peak was considered to match only if its t_R fit within ± 0.2% to the expected t_R and if 10 or more other peaks could be matched with the same precision. 0.5 ml fraction were collected in polypropylene tubes containing protease inhibitor cocktail solution and analyzed by SDS-PAGE.

The immunoaffinity protocols were adapted from common procedures (for a review, Warren, 1994; Schneider et al., 1982). The rabbit polyclonal antiserum (pAb) raised against the synthetic peptide ARLWTEYFVI and the mouse monoclonal antibody (mAb) raised against the denaturating gel-extracted purified protein were provided by T. Nelson (Las, NIH, Bethesda, MD, USA). In the first purification attempt, the synthetic peptide was immobilized on an activated CNBr-agarose gel (Sigma, St. Louis, MO, USA), following the manufacturer's protocol (see also Husten and Eipper, 1994). Resin was then packed in a 2 ml polypropylene column (Bio-Rad) and 1 ml aliquots of pAb were loaded. The flow rate was 1 ml/min and the temperature 8°C. After washing the stationary phase with 10 column volumes of tris buffer saline (TBS) and 5 column volumes of 1M NaCl in TBS to reduce non-specific binding, the retained fraction was eluted with diluted HCl (pH 2.7) and collected in 0.4 M tris-containing tubes (final pH 7.4). The column was then washed with HCl (pH 0.5) and regenerated with water, and another aliquot of pAb was loaded. The fractions were tested by measuring the absorbance at 280 nm (A280, total protein content) and by dot-blot immunostaining, using a secondary goat anti-rabbit polyclonal antibody conjugated with alkaline phosphatase (Pierce, Rockford, IL, USA), and stained with nitro-blue tetrazolium (NBT) and 5-bromo-4-chloro-3-indoyl phosphate (BCIP) at pH 9.5 (trisCl). The active fraction (purified polyclonal antibody, ppAb) was immobilized on a protein A-sepharose matrix (UltraLink, 60 µm particle size, 100 nm pore size; Sigma, St. Louis, MO, USA), according to the manufacturer's procedure (see also Warren at al., 1994), and six 2 ml plastic column were gravity-packed at 4°C. Samples (2 ml) of squid optic lobe supernatant prepared as above were applied to the columns at a flow rate of 1 ml/min. After washing with 10 column volumes of water, the retained fractions were eluted by 1M NaCl in 0.1 M glyCl (pH 2.7), and collected in tubes containing 0.4 M Tris (final pH 7.4). The fractions were tested for A280, and the retained peaks were analyzed by SDS-PAGE.

In the second immunoaffinity protocol, the mAb was immobilized on protein A with a similar procedure (Sagoo et al., 1994; Yloenen et al., 1994), and squid optic lobe supernatant was prepared as above. Rabbit brain cortex (Peel Freeze, NY, USA) was homogenized by sonication at 0°C in loading buffer with protease inhibitor cocktail. After high-speed centrifugation (100000 x g), the pellets were resuspended in 6M urea and rocked at room temperature for 30 minutes. After a second ultracentrifugation, the supernatant was collected. Rabbit and squid supernatant were loaded on the mAb column, at a flow rate of 2 ml/min (washing buffer: 10 column volumes of 0.1 M borate buffer, pH 8.2; eluting buffer, after equilibration with water: diluted KOH, pH 12.5, with 0.5 M monobasic potassium phosphate in collecting tubes). The retained fractions were desalted by concentration on Centricon C10 and dilution with protease inhibitor cocktail, and loaded on a GTP-agarose affinity column (V=1.5 ml; Sigma, St. Louis, MO, USA). The columns were washed at 4°C and 1 ml/min with 10 column volumes of water, and then a linear gradient to 4.2 M NaCl was applied over 20 minutes, followed by 4.2 NaCl for 10 minutes. 2 ml fractions were collected, tested for A280, and analyzed by SDS-PAGE.

The preparative protocol was a modification of the short protocol described above. Two batches of 400 squid optic lobes were high-speed homogenized (Polytron, Switzerland, setting 5) at 4°C in 150 ml of the same above buffer with protease inhibitors, and sonicated in ice at maximum power 4 times 30 seconds. An aliquot of the homogenate (*crude extract, CE*) was saved for protein quantitation and SDS-PAGE, while the remaining was centrifuged at 150000 x g (Becmkan 70Ti rotor) for 2 hours at 4°C. The pellet was rehomogenized in 150 ml of buffer and recentrifuged in the same conditions. An aliquot of the combined supernatant (*ultracentrifuged, UC*) was saved for analysis. The solution was passed through 8 Centriprep C50 (two loads were necessary), which also eliminated the residual particles. The low molecular weight fraction (from which an aliquot, *ultrafiltered or UF*, was saved) was applied on a low-pressure, preparative weak anion-exchange glass column (Pharmacia, 50 x 3 cm with polystyrene flow corrector; fast-flow DEAE-sepharose, Sigma), water-jacked at 8°C. The flow rate was maintained at 1.5 ml/min along the loading and the washing (water, 12 hours). The retained fraction was eluted isocratically at 6 ml/min with 1M potassium acetate buffer (pH 7.4). After saving an aliquot (*DEAE-retained, DE*) the sample was extensively desalted through Centriprep C10 (final ionic strength < 30 mM), using protease inhibitors as a diluting solution. The resulting volume (300 ml) was loaded on a medium-pressure dye-affinity column (Pharmacia, 25 x 2 cm with flow corrector; fast-flow Cibacrom Blue-agarose, Sigma), always at 8°C. The flow rate was 2.5 ml/min, and after 7 hours of water washing, the column was eluted with 1 M potassium acetate (pH 7.4). An aliquot was saved (*Cibacrom Blue-retained, CB*), and the sample was desalted and concentrated in Centricon 10 down to 10 ml. This volume (*Injected sample, In*) was divided into 5 loads, and injected with a 2 ml loop on a semi-preparative AX-300 anion exchange HPLC column (1 x 25 cm, 30 nm gel particle diameter; Thomson, Springfield, VA, USA), equipped with a AX-300 guardagel precolumn (Thomson, 2.5 x 1 cm). The column was eluted at 2 ml/min and 8°C with a linear gradient of 0-0.6 M potassium acetate buffer (pH 7.4) for 20 minutes followed by 0.6 M buffer for 30 minutes and 1 M buffer for 10 minutes. The guardagel precolumn was changed every two preparations. The chromatograms were analyzed by computer-assisted peak matching as above. The fractions were collected every 0.5 minutes, diluted with protease inhibitors, and stored at -80°C before SDS-PAGE analysis. In some preparations, the samples containing cp20 were pooled together (*AX-300 retained, AX*), concentrated to 0.5 ml and reinjected with a 0.5 ml loop on the same system, to increase the purity.

Some products were analyzed by reverse phase chromatography (RP-HPLC). An aliquot of 0.1 ml was injected in loop onto a C18 column (0.25 x 25 cm, 30 nm gel particle diameter; Synchrom, Lafayette, IN, USA). The column was developed at 0.5 ml/min with a 20-100% linear gradient of acetonitrile (+0.1% trifluoroacetic acid, TFA) in water (+ 0.1% TFA) over 90 minutes, followed by 100% acetonitrile (+ 0.1% TFA) for 30 minutes. 3 minute fractions were collected, frozen, lyophilized, suspended in TBS and analyzed by SDS-PAGE (*reverse phase samples, RP*).

Purified squid cp20 was stored in potassium acetate buffer with protease inhibitors in small aliquots (50 μl, about 0.5 μg), and was stable for several months at -80°C.

Results and Discussion

With the *short protocol* based on a single HPLC step, 50 to 100 ng of squid cp20 could be extracted from 25 animals, with a final purity ranging between 40 and 60%. While this yield allowed some preliminary biochemical characterization (Nelson et al., 1994; Nelson and Alkon, 1995; Nelson et al., 1996a) it was not enough to attempt a structural study.

A polyclonal and a monoclonal antibodies were raised against cp20 and the partial sequence ARLWTEYFVI, respectively. The polyclonal antiserum (pAb) was immunopurified by antigen-affinity chromatography (fig. 7), and the resulting product (ppAb) showed an improved specificity on western blot.

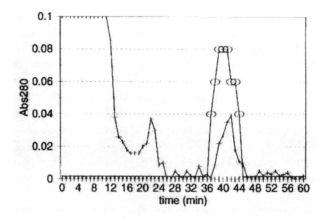

Figure 7. *Elution of pAb on a cp20 peptide-based column. Circles: immunoreactivity (arbitrary OD units); crosses: total protein.*

The relative simplicity of the immunoaffinity purification protocols (Warren, 1994; Warren et al., 1994) allowed the parallel use of several ppAb-protein A columns. Six small-scale samples of squid optic lobe crude extract supernatant were fractionated. A retained peak was consistently recovered (figure 8), and the fractions were analyzed by SDS-PAGE.

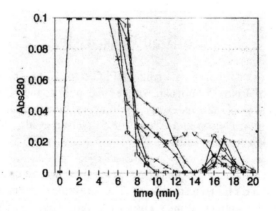

Figure 8. *Elution of squid optic lobe protein crude extract supernatant on a ppAb-based column from six different experiments.*

A mixture of proteins was observed in the retained fraction, with a major component at 31 kD and no visible bands in the 20 kD range. The preparations were thus discarded, and another method attempted.

The mAb showed a higher specificity on western blot of squid optic lobe and Hermissenda CNS crude extracts, and it also individuated a 20 kD band in the rabbit brain cortex. This rabbit protein had been recognized as cp20 (Nelson et al., 1994), was mainly compartmentalized in the cytosolic membrane and changed upon associative conditioning (Nelson et al., 1991). A urea-based extraction of rabbit brain cortex proteins was set up. The use of urea instead of common detergents is justified by the fact that the mAb was raised against the denatured protein (Nelson et al., 1994), and urea is easily eliminated by desalting procedures. Squid and rabbit extracts were loaded on the mAb column (fig. 9), and two sharp retained peaks were observed.

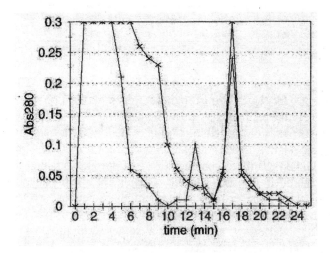

Figure 9. *Elution of squid optic lobe (X) and rabbit brain cortex (+) protein extracts on a mAb-based column.*

An SDS-PAGE analysis of the peaks revealed high protein content and extremely low purity. A second affinity chromatographic step was performed, using the GTP-binding properties of cp20 (Nelson et al., 1990, 1991, 1994). The samples from rabbit and squid were loaded on a GTP-agarose stationary phase (figure 10), and the elution required a very high ionic strength; the samples were therefore desalted by ultrafiltration before analysis.

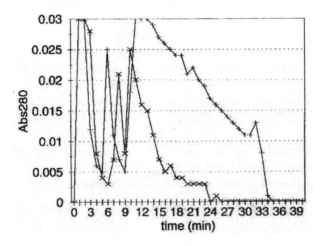

Figure 10. *Elution of the retained peaks from the chromatogram of fig. 9 on a GTP-agarose column.* X, squid; +, rabbit.

None of the fractions contained 20 or 40 kD band on SDS-PAGE. One of the possible reasons for the failure of immunoaffinity purification was the relatively low specificity of the antibodies. Very particular conditions were necessary in order to obtain single bands on western blot (Kim et al., 1995) for both ppAb and mAb. In addition, the relative amount of cp20 in the tissues is quite low, and the column capacity is limited; in these conditions, a low specific binding affinity can be completely inhibited by non-specific binding of more abundant proteins. Furthermore, due to the extreme dilution of the solutions, a large amount of membrane absorption was observed on ultrafiltration devices (see also Suelter and DeLuca, 1983; Nelson et al., 1994). A larger-scale protocol was therefore necessary to limit this inconvenience.

The five-step protocol described at the end of the experimental section combined some elements of the anion-exchange procedure with affinity purification protocols. Some technical improvements were necessary for scaling up the purification. 800 squid optic lobes were divided into two batches and high-speed homogenized, followed by sonication in ice. The ultracentrifugation step yielded a cytosolic fraction slightly contaminated by low-density particles (possibly a phospholipidic fraction). The double molecular weight cut-off (Nelson et al., 1994, Nelson and Alkon, 1995), which caused major loss of cp20, was substituted by a single 50 kD ultrafiltration step. This was performed by special devices originally designed for cell suspensions (Amicon Centriprep), which yielded a perfectly transparent, protein-rich fraction, free of high-molecular weight macromolecules (nucleic acids, proteoglycanes, etc.). This solution could not be frozen, because some low-solubility proteins irreversibly precipitated at this stage, dramatically decreasing the final yield of cp20. The large sample (over 300 ml containing more than 2g of proteins) was processed by chromatography on a large scale, preparative, weak anion-exchange resin (fig. 11). This step allowed an over night fractionation, with a noticeable increase in the concentration of low-isoelectric point proteins (including cp20). Furthermore, many of the non-proteic components were eliminated.

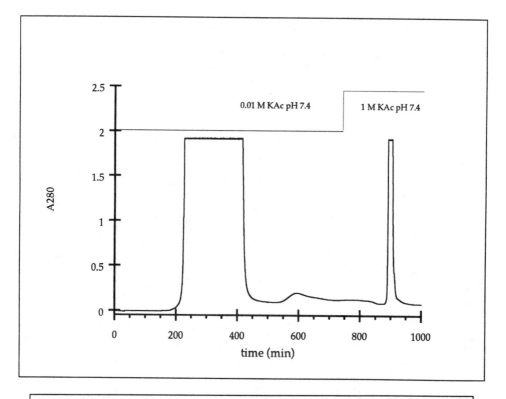

Figure 11. *Overnight elution of a crude extract supernatant from 400 squid optic lobes on a preparative weak anion-exchange low-pressure chromatographic column.*

In some preparations, NaF was added to the mobile phase to prevent phosphatase activity, but this did not increase the final yield, and damaged the stainless steel HPLC tubing. In contrast, it was important to add protease inhibitors to the retained fraction upon collection and successive desalting and concentration.

The resulting large volume (280 ml) was loaded by continuous flow on a Cibacrom Blue dye-affinity column. The void volume was passed through a second time, and the entire loading operation (tot. 6 column volumes) was carried out at a low flow rate to allow a longer interaction time with the stationary phase. Cibacrom Blue is an affinity medium with a broad range of protein binding activity. The stationary phase active compound is structurally related to ATP/GTP-like molecules; although the interaction mechanism is not known, this compound retains other G-proteins of the cp20-ARF superfamily (Lanznaster and Croteau, 1991; Su et al., 1991; De Matteis et al., 1993, and references therein). The retained fraction was eluted isocratically (fig. 12), diluted with protease inhibitors, desalted and concentrated.

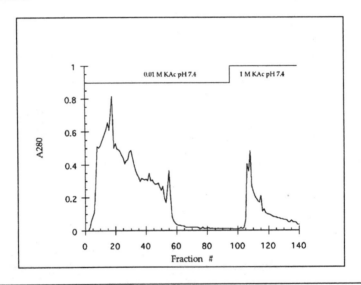

Figure 12. *Chromatogram of the retained peak from fig. 11 eluted on a Cibacrom Blue column. Fractions were collected every 5 minutes (12.5 ml).*

Cp20 is probably eluted at the beginning of the retained peak (Nelson, personal communication), and experiments are in progress to ensure a more efficient fractionation by a linear gradient elution. At this stage, cp20 represented 1-3% of the total proteins, i.e. it was over an order of magnitude more concentrated than in the short protocol before injection in HPLC.

The solution was injected in batches onto the AX-HPLC. Before such a last fractionation, it was not possible to freeze the samples because precipitation occurred. It was therefore extremely important to carry out all the three-day operations at 0-4°C and in the presence of protease inhibitor cocktail. The AX-HPLC step yielded the chromatogram in fig. 13, with a characteristic relatively large peak at 31.5 minutes on the tail of the major peak at 25.5 minutes.

Temperature did not affect the chromatographic pattern, and it was thereby kept constant at 8°C. In contrast, the composition of the potassium acetate buffer was critical; if the pH equilibration was carried out by an acid different from acetic acid (e.g., chloridric acid), a different chromatographic shape was observed.

The chromatographic pattern was extremely reproducible, among several injections in the same preparation and among different preparations. The introduction of a guardagel precolumn did not sensibly affect the retention time, but increased the column efficiency upon repeated injections. The cp20-containing peak was identified at 31.5 minutes by computer-assisted pattern matching, with a linear correlation factor over 99.9% (fig. 14). Almost all the 18 distinguishable peaks fit the correlation, and the void time was virtually identical to that of the reference chromatogram.

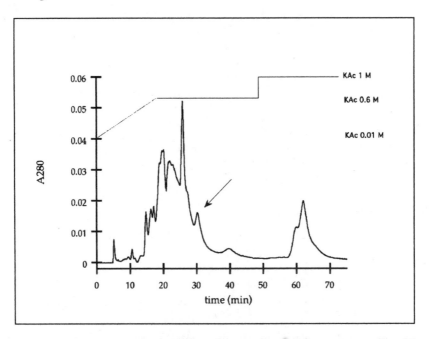

Figure 13. *Semi-preparative AX-300 HPLC elution of the retained fraction from the chromatogram of fig. 12. The arrow indicates the cp20-containing peak.*

The candidate peak was in a relatively empty chromatographic region, and the correlation was unique, thus allowing an unequivocal identification.

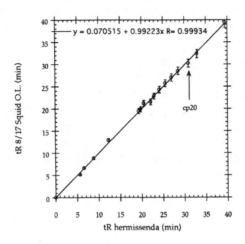

Cp20 eluted at 0.6 M KAc, as previously reported (Nelson and Alkon, 1995). The cp20 peak was on the tail of a previous large peak (at 25.5 minutes, fig. 15). Half minute fractions were collected, pooled by two, and analyzed by SDS-PAGE.

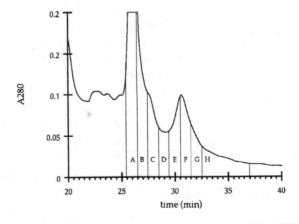

Figure 15. *Enlargement of the cp20-containing peak of fig. 14 and definition of collected fractions (H is a 4.5 minute tail).*

Fractions eluted prior to 25.5 minutes were not analyzed, while the last fraction, between 32.5 and 37.0 minutes was concentrated and analyzed, but showed no protein bands. The silver stained gel of fractions A-H is reported in fig. 16.

A concentrated protein is visible in fraction F, with a tail in G. The molecular weight (about 18 kD) is low with respect to the expected one. A band just above the 21.5 kD molecular weight marker is visible in fraction D, at a retention time previously correlated with non-phosphorylated cp20 (Kim et al., 1995; Nelson and Alkon, 1995). The 18 kD band at 31.5 minutes and the 22 kD band at 28.5 minutes were blotted, partially digested, and showed overlapping sequences (Nelson et al., 1996b). This finding indicates that the 18 kD band is actually a degradation product of the 22 kD band.

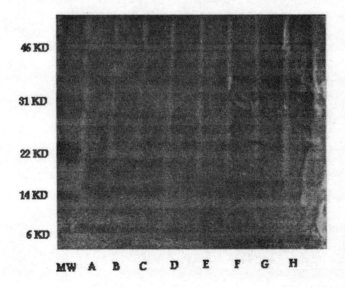

Figure 16. *Silver-stained SDS-PAGE of the fractions defined in fig. 15.*

When the 22 kD band was repurified as described below, it was identified as the entire cp20, and when it was degraded under controlled conditions (in water at room temperature for 6 hours), a major band at 18 kD appeared on SDS-PAGE.

Fraction B, corresponding to a shoulder in the chromatogram of figures 14 and 15, was linearly correlated with another protein in the Hermissenda references chromatogram, cp27 (Nelson et al., 1990; Nelson and Alkon, 1991), which also changes upon classical conditioning. Moreover, it is possible to notice that all major impurities of the cp20 band in fraction D are constituted by proteins from the tail of the large peaks at 25-27 minutes. In fact, fraction D does not even elute as a peak in the chromatogram. It was then possible to further purify fraction D by reinjection on the same HPLC system. The lower load allowed a much better chromatographic separation (fig. 17), and a pure single peak at the same retention time (28.5 minutes) could be collected.

The SDS-PAGE analysis of the cp20 reinjected sample showed a single band upon silver staining, positioned between the molecular weight markers at 21 and 31 kD (fig. 18). The recovery was low, due to non-specific absorption of the pure protein on the concentrator membranes. The yield was increased by BSA pretreatment of the ultrafiltration apparatus, but in this case a major BSA impurity (66 kD) became visible on the silver stained gel.

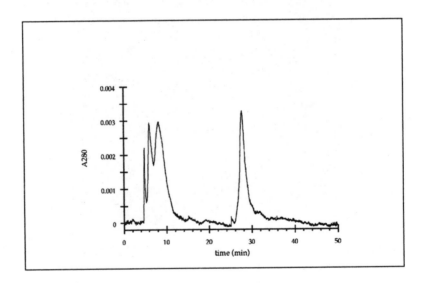

Figure 17. *AX-300 HPLC chromatogram of reinjected fraction D from fig. 15. The elution gradient conditions are identical to the chromatogram of fig. 15.*

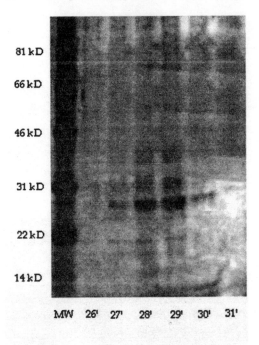

Figure 18. *Silver-stained SDS-PAGE from the chromatogram of fig. 17. Graphic file available by anonymous ftp at linus.ninds.nih.gov/mcid/img.*

The use of a 30 kD cut-off in the previously adopted protocols (Nelson et al., 1994, and the short protocol described in this chapter) had caused the elimination of the 22 kD band, and the 18 kD fragment was the only one that could be observed.

The fractions containing repurified cp20 and the partially purified 18 kD fragment were stored at -80°C with protease inhibitor cocktail. They were then analyzed by reverse-phase (RP) HPLC, as described in a previous test (Nelson et al., 1994). Small aliquots were thawed and immediately injected on the C18 column to provide an optimized separation. Thanks to the desalting property of the reverse-phase chromatographic system, the fraction could be collected, lyophilized and entirely loaded on SDS-PAGE.

The chromatographic elution of the fraction containing the intact cp20 is reported in fig. 19. A blank chromatogram obtained by injection of a mixture of the potassium acetate buffer and the protease inhibitor cocktail was subtracted.

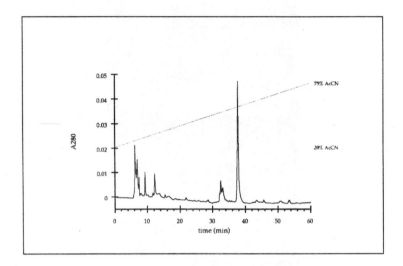

Figure 19. *C18 reverse-phase chromatographic elution of the intact cp20 (AX 300 HPLC reinjected fraction D, figures 15-18).*

The A_{280} trace shows a single peak at the expected t_R (38 minutes), in agreement with the high purity observed on gel. The following SDS-PAGE confirmed the molecular weight. The peaks at 33 minutes contained BSA from ultrafiltration membranes, as proved by the SDS-PAGE molecular weight and by the injection of a solution obtained passing water through the concentrator devices.

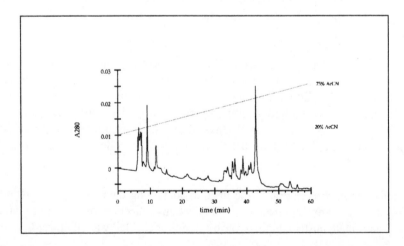

Figure 20. *C18 Reverse-phase chromatographic elution of the major peak from fragmented cp20 (fraction F in fig. 15-16).*

The fraction containing the partially purified cp20 fragment (peak F in figures 14-16) yielded the RP chromatogram reported in fig. 20.

The major peak, at 42.5 minutes, was the 18 kD protein observed in the SDS-PAGE. As expected, its polarity (and thus RP-HPLC retention time), was similar to that of the entire cp20. Among the other peaks, BSA is still visible, while the intact cp20 elutes at 37.5 minutes, as in the previous chromatogram. The remaining peaks were matched with the impurities present in fraction F (fig. 16) by SDS-PAGE analysis.

The summarizing data of yield and purity of the cp20 preparation from squid optic lobe are reported in tab. 1.

	TotProt	TotCp20	Purity %	StepRec	StepPF	TotRec	TotPF
CE	6000000	600	0,01	1	1	1	1
UC	4000000	300	0,0075	0,5	0,375	0,5	0,375
UF	1000000	200	0,02	0,67	1,79	0,33	0,67
DE	250000	100	0,04	0,5	1	0,17	0,67
CB	50000	40	0,08	0,4	0,8	0,067	0.53
IN	40000	35	0,09	0,87	0,98	0,058	0.52
AX	40	15	40	0,43	191,1	0,025	100
AX2	4,4	4	90	0,37	0,83	0,007	60
RP	1	1	>99	0,25	0,28	0,002	16,67

Table 1. *Purification data. Vertical titles (samples): CE, crude extract; UC, ultracentrifuged; UF, ultrafiltered; DE, DEAE-retained; CB, Cibacrom Blue-retained; IN, HPLC-injected (after Centricon C-10 passage); AX, AX300-retained; AX2, second passage through; RP, reverse phase-purified. Horizontal titles (parameters): TotProt, total protein amount (in µg); TotCp20, total cp20 amount (22 kD band + 18 kD band), as estimated by SDS-PAGE (in µg); Purity %, TotCp20/TotProt*100; StepRec, step recovery, ratio between the TotCp20 values after and before the step; StepPF, step purification factor, product of the step recovery times the ratio between the purity values after and before the step; TotRec and TotPF, total recovery and total purification factor, respectively, are calculated with a formula analogous to that of the step values, referring to the beginning of the purification (CE step) instead of to before the step.*

It is clear that the crucial purification step is the AX passage. All the steps prior to the anion-exchange HPLC are useful to obtain a cp20-rich solution, but are not highly efficient in terms of purification parameters such as purity and recovery, as indicated by the step purification factor and the total purification factor, which represent the efficacy of the passage. The convenience of these steps is testified by the success of the preparation with respect to the short protocol described at the beginning of the chapter; the AX column is loaded with a saturating amount of protein, but the concentration of cp20 is one order of magnitude higher than in previous protocols at the corresponding passages. The Cibacrom Blue passage could be optimized to increase the purification factor (experiments in progress). The purification steps after AX, in contrast, are performed to achieve a purity suitable for analysis and molecular biology, but cause a remarkable loss of cp20. Once again, when the amount of protein is too low, unspecific absorption on membranes, tube walls etc. becomes a determining factor.

The reverse-phase purified cp20 and the AX-reinjected purified protein were analyzed by ion-spray mass spectrometry and Fourier-transform infrared spectroscopy, but both attempts failed because of the small size of the samples.

The gels with purified cp20 and its major fragment were then electroblotted on a polyvinyl difluoride membrane and subjected to tryptic digestion for sequencing. In particular, the 18 kD band could be selectively blotted, probably because of its low molecular weight and isoelectric point, to yield an almost pure band on the membrane, as tested by Poinceau S staining. The crude digested extract was separated by capillary electrophoresis, and fourteen peptide fragments (118 aminoacid total) determined to be pure by mass spectrometry were sequenced (fig. 21). Degenerated oligonucleotides based on the peptide sequences were used to amplify complement-deoxyribonucleic acid prepared from squid optic lobe polyadenosine messenger ribonucleic acid using the reverse transcriptase polymerase chain reaction. The 247 products were cloned into a commercial system and used to transform an XL cell line (Nelson et al., 1996b). All clones were resequenced twice by independent operators, and one was found that contained 13 of the 14 peptides (fig. 21).

A 52 base pair oligonucleotide based on the clone sequence was labeled with T4 kinase and used to screen a squid optic lobe genetic library. One 1.2 kbase phage clone was found and subcloned into a commercial system. The 722 base pair product was cloned into a commercial vector and subcloned to add a 2 kD N-terminal oligohistidine domain. The resulting product was used to transform a line of *Escherichia coli* cells (Nelson et al., 1996b). The resulting protein was then expressed and sequenced as described in the next chapter.

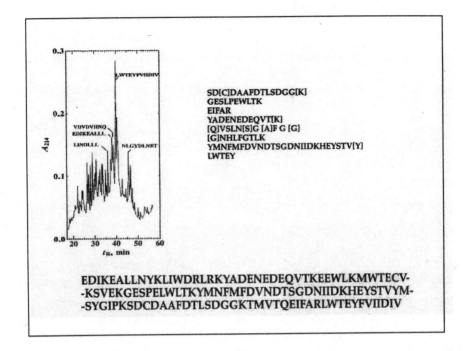

SD[C]DAAFDTLSDGG[K]
GESLPEWLTK
EIFAR
YADENEDEQVT[K]
[Q]VSLN[S]G [A]F G [G]
[G]NHLFGTLK
YMNFMFDVNDTSGDNIIDKHEYSTV[Y]
LWTEY

EDIKEALLNYKLIWDRLRKYADENEDEQVTKEEWLKMWTECV-
-KSVEKGESPELWLTKYMNFMFDVNDTSGDNIIDKHEYSTVYM-
-SYGIPKSDCDAAFDTLSDGGKTMVTQEIFARLWTEYFVIIDIV

Figure 21. *Sequence of cp20 18 kD major fragment's tryptic peptides fractionated on reverse-phase HPLC (upper left panel), and of intact cp20 tryptic peptides separated by capillary electrophoresis (upper right panel). Cp20 clone obtained from the fragments EDIKEALL and EIFARLWTEY (lower panel).*

CHAPTER 2

Purification and preliminary characterization of cloned cp20

The cloning of squid cp20 allowed the characterization of its aminoacidic sequence. The information on the primary structure was yet not complete because post-translational modifications were missing. However, the study of the aminoacidic sequence made it possible to evaluate the consensus sites for physiologically meaningful common post-translational modifications. Cp20 proved to be a potential signaling protein with an unusually high capacity for cellular transduction. For example, two calcium-binding domains were present, in addition to the expected PKC-phosphorylation site and several membrane-anchoring consensus sequences.

The chemical and physical properties of the protein as calculated by the primary structure (e.g. MW and isoelectric point) were in excellent agreement with the experimental data previously measured on the squid optic lobe's natural protein.

In order to study experimentally the biochemical characteristics of cp20, a bacterial expression system was set up. This provided a large amount of cp20-rich tissue from which relatively high quantities of protein could be extracted and isolated. In particular, the oligo-histidine tail fused with the cloned protein showed a high affinity for Ni^{2+}. This property allowed a rapid affinity purification based on high-performance metal-chelated chromatography (IMAC-HPLC) after the usual passages of homogenization and ultracentrifugation. The large-scale preparation (from 20 liters of bacteria) yielded several hundred micrograms of pure cp20, which allowed the protein to be more thoroughly characterized. This chapter describes the biochemical and physiological properties of the protein, while the structural and spectroscopical studies are reported in the next chapter.

A new polyclonal antibody was raised against a portion of the sequence which was much larger than that used for the previous antibody (chapter 1). This new antibody showed excellent specificity and good sensitivity. It could be used to analyze the purification products, and recognized both the cloned protein and the natural squid cp20 at the expected molecular weights by western blot analysis.

The purified cp20 showed electrophysiological activity identical to that of the natural protein, with an inhibition of the calcium-dependent potassium channels and a consequential decrease in the related currents. In addition, the fact that cp20 can act on the electrochemical properties of cells from various species (Hermissenda, rabbit, human) strongly suggests that this protein is well conserved through evolution.

Cloned cp20 was also phosphorylated by the α-isozyme of PKC and showed both GTP-binding and GTPase activities, like the squid optic lobe protein. The identity with the natural cp20 was further probed by reverse-phase chromatography, which showed that the two proteins have very similar retention times.

Finally, cp20 calcium binding was investigated. As expected from the primary structure, two moles of calcium could be complexated by one mole of protein, as assessed by overlay blot. In addition, dialysis experiments and Scatchard analysis indicated a binding constant ranging around 40 nM for the high-affinity site.

The calcium binding properties together with the neuron hyperpolarization activity suggests a new name for cp20: Calexcitin.

Materials and Methods

Protein sequence analyses were performed with databases available on the internet (Expasy, Switzerland); the search for post-translational modifications and physical-chemical parameters were performed with software on related web sites (SwissProt and ProtParam, respectively).

The 26-amino acid oligopeptide YMNFMFDVNDTSGDNIIDKHEYSTVY, homologous to a central position on cp20 primary structure, was purchased (Genosys, NJ, USA), KLH-conjugated and injected into three rabbits according to the manufacturer's protocol for polyclonal antibody production. The 8th bleed (100th day after the first injection) was tested and used as active antiserum. Aliquots of 1 ml were lyophilized with 0.03% NaN_3 and frozen at -80°C for extended storage. In western blot experiments, Novex gels were transferred in a semidry electroblotting apparatus (Bio-Rad, Richmond, CA, USA) according to the manufacturer's protocol. Briefly, a pH gradient (8.7 to 10.7) was obtained by stacking unmodified cellulose sheets soaked in 20% methanol/water solutions of γ-aminocaprylic acid between the anode and the cathode. The gel and the nitrocellulose membrane (Pierce, Rockford, IL, USA) were sandwiched on top of the pH 8.7 layer, with the nitrocellulose membrane facing the high pH side. After blotting at constant current, the gel was either discarded or silver stained for residual proteins, and the membrane was washed 10 minutes with water and dried. After rehydrating in TBS, the membrane was incubated for an hour in polyclonal antiserum (1:1000 in TBS), washed in 0.1% triton-X/TBS (TBST) 3x10 minutes, and incubated for an hour in 1:5000 alkaline phosphatase-conjugated goat-antirabbit polyclonal antibody (Sigma, St. Louis, MO, USA). After a second washing cycle in TBST, the membrane was developed with NBT/PCIB at pH 9.5 and 2 mM $MgCl_2$. Upon appearance of the bands, the reaction was stopped by rinsing with water. All steps were performed at room temperature with gentle rocking. The membrane was then dried and scanned with the same imaging system used for gels. The antibody was tested by two blanks, i.e. prebleed and preabsorbed antibodies. The prebleed is the polyserum obtained from the rabbit before the injection of the antigenic peptide, while the preabsorbed is the active antibody preincubated with 1 mg/ml antigen for one hour at room temperature prior to use. Proportional protein amounts were usually loaded in silver stain and proportional volumes in western blot. The detection limit for cp20 was 25 ng in western blot and 50 ng in silver stain.

Squid axons were purchased from Calamari Inc. (Woods Hole, MA, USA), stored at -80°C and treated like the optic lobes (Chapter 1) to prepare a crude extract fraction.

Escherichia coli cells genetically transformed with cp20 were provided by Tomas Nelson (Las, NIH, Bethesda, MD, USA), and conserved in water/ethylene glycol at -80°C. The cells were stripped in sterile conditions on a Louria-Bertani (LB)-ampicillin plate (Sigma, St. Louis, MO, USA), and from this plate immediately reinoculated onto a second plate. The two plates were incubated at 31°C for two days, and one of the two developed spot-colonies. This plate was parafilmed and stored at 8°C, and a new plate was restripped from the previous one every two weeks.

A bacterial stock solution was prepared by stirring one colony in 25 ml of LB medium containing 100 µg/ml ampicillin. The suspension was shaken vigorously at 37°C with loose cap to allow aeration, and the bacterial growth checked by the absorbance of a 1.5 ml-aliquot at 600 nm (A600). The incubation was stopped at early log-phase (A600=0.4) and stored tight-cap at 4°C.

20 liters of ampicillin-modified LB-broth were incubated with 1 ml of stock solution each at 37°C, with continuous stirring and aeration. A600 was checked every 30 minutes, and isopropyl-thioglucopiranose (IPTG) was added (1mM final) at mid-log phase (A600=0.6). The incubation then continued for three hours, with a final A600 of 1. The cells were centrifuged (5000 x g, Sorvall variable angle rotor), resuspended in 90 ml of TBS containing protease inhibitor cocktail and imidazole (Iz) 0.5 mM and stored at -80°C. Ampicillin, IPTG and protease inhibitor cocktail (prepared as described in the Materials and Methods of chapter 1) were aliquoted in 1000x mother solutions, after sterilizing by 0.2 µm filtration, and conserved at -80°C.

Upon partial thawing, the cells were homogenized at 0°C by high-power sonication (an aliquot, *CE* or *crude extract*, was saved for protein analysis and SDS-PAGE) and ultracentrifuged for one hour at 4°C and 100000 x g (Beckman, rotor Ti70). The pellets were resuspended in 30 ml of the same buffer and recentrifuged. Pooled supernatants (*ultracentrifuged, UC*, aliquot) were divided in two batches and loaded on a His-Tag column (15 ml; Novagen, NY, USA; water-jacked at 8°C), previously charged with an excess of $NiSO_4$ or $NiCl_2$[1]. The column was washed with at least 10 volumes each of binding buffer (20 mM trisCl, 0.5 M NaCl and 0.5 mM Iz, pH 7.9) and washing buffer (same with 25 mM Iz), at a flow rate of 2.5 ml/min. Upon stabilization of the baseline, the column was developed with eluting buffer (same as above, with 0.5 M Iz). 10 ml fractions were collected and immediately diluted with 5 ml of 3x protease inhibitor cocktail; a 50 µl aliquot (*his-tag, His*) was saved for SDS-PAGE analysis, and the fractions were stored at - 80°C.

Fractions containing cp20 according to SDS-PAGE and western blot were pooled together (about 150 ml), desalted and concentrated at 8°C down to 10 ml (final *C10* fraction). Some precipitation occurred, and a medium-speed centrifugation (Sorvall S34 rotor, 4°C) was performed. Pellets were stored at -80°C (*precipitate pellet, PP* fraction), and the supernatant was used in the following step of the purification.

Immobilized metal-affinity chelation high-performance liquid chromatography (IMAC-HPLC) was performed with a MC-POROS system (PerSeptive Biosystems, Cambridge, MA, USA), at room temperature and with a flow rate of 5 ml/min; the perfusion column (100 x 4.6 mm, 20 µm particle diameter) was charged with 25 column volumes of 0.2 M Ni^{2+} according to the manufacturer's recommendations, and equilibrated in binding buffer. The column was tested by injecting two solutions containing 20 and 100 µg of horse heart myoglobin, respectively, and developing with 10 column volumes of 0.5 mM imidazole followed by 5 column volume of 50 mM imidazole. The commercial protein eluted in the retained fraction. The His-Tag fractions containing cp20 were loaded onto the column after adding NaCl and imidazole to a final concentration of 0.5 M and 0.5 mM, respectively. A 2 ml loop was used, and five repeated runs were performed in each preparation. The loaded column was washed in succession with binding buffer until baseline stabilization, and developed with a linear gradient from 0.5 mM to 500 mM of imidazole in 15 minutes. 0.5 min fractions were collected, diluted with protease inhibitors and stored at -80°C after saving an aliquot (*IMAC-retained, IM*) for analysis. The column was stripped in 100 mM EDTA for 3 min, at a flow rate of 3 ml/min, equilibrated with water, and recharged before each run. The column was regenerated with NaOH/HCl cycles as suggested by the manufacturer after every preparation, and stored at 8°C.

The fractions containing cp20 were pooled together, desalted with protease inhibitor, concentrated to 0.5 mg/ml and aliquoted (50 µl) for prolonged storage at -80°C.

---

[1] Both salts are colored (blue and green at neutral pH, respectively), and the coloration of the white resin can be followed during the metal load.

Electrophysiological, Ca^{2+} binding and PKC phosphorylation experiments were performed as described (Nelson et al., 1996b), while RP-HPLC and AX-HPLC experiments were performed as reported in Chapter 1.

Results and Discussion

The cp20 cDNA sequence shows close homology to several calcium-binding proteins, particularly scp1. There are also regions with some homology to GTP-binding proteins of the adenosine ribosylation factor (ARF) family, such as sar1p. The sequence data of squid cp20/calexcitin (GenBank accession number U49390) are reported in fig. 22.

Northern blot analysis showed two bands (approx. 9.99 and 8.17 kbases). Reverse transcriptase polymerase chain reaction experiments excluded the presence of alternative splicing on the coding region. The 3'-untranslated region contained long stretches of poly-(AT), a common finding in squid mRNAs.

The complete aminoacidic sequence of the cloned protein (with the fused oligohistidine tail) is shown in fig. 23.

The sequence -9 to -1 is the consensus site for a specific protease, activated factor ten, and has been introduced in order to cleave the oligohistidine chain after the purification (as reported at the end of chapter 3), if necessary. The calculated isoelectric point (4.97) agrees with the pI measured by isoelectric focusing (5.3) and with that observed for purified native squid and Hermissenda cp20 (pI = 5.2) on 2D SDS-PAGE (Nelson et al., 1994 and 1996b). Also the calculated molecular weight (24434.1 with histidine tail, 21927 without) is in agreement with the experimental data (SDS-PAGE and size-exclusion chromatography, Nelson et al., 1994).

The sequence contains a PKC phosphorylation site in position 61-63. This finding was clearly expected (cp20 was originally discovered as a phosphoprotein), but suggests that a threonine rather than a serine is the phosphorylated residue. *In vivo*, PKC exhibits a preference for the phosphorylation of serine or threonine found close to a C-terminal basic residue (see e.g. Lester, 1992).

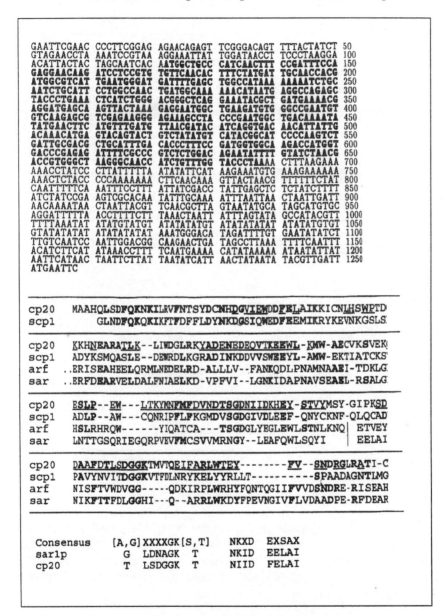

```
GAATTCGAAC CCCTTCGGAG AGAACAGAGT TCGGGACAGT TTTACTATCT 50
GTAGAACCTA AAATCCGTAA AGGAAATTAT TGGATAACCT TCCCTAAGGA 100
ACATTACTAC TAGCAATCAC AATGGCTGCC CATCAACTTT CCGATTTCCA 150
GAGGAACAAG ATCCTCCGTG TGTTCAACAC TTTCTATGAT TGCAACCACG 200
ATGGCGTCAT TGAATGGGAT GATTTTGAGC TGGCCATAAA AAAAATCTGC 250
AATCTGCATT CCTGGCCAAC TGATGGCAAA AAACATAATG AGGCCAGAGC 300
TACCCTGAAA CTCATCTGGG ACGGGCTCAG GAAATACGCT GATGAAAACG 350
AGGATGAGCA AGTTACTAAA GAGGAATGGC TGAAGATGTG GGCCGAATGT 400
GTCAAGAGCG TCGAGAAGGG AGAAAGCCTA CCCGAATGGC TGACAAAATA 450
TATGAACTTC ATGTTTGATG TTAACGATAC ATCAGGTGAC AACATTATTG 500
ACAAACATGA GTACAGTACT GTCTATATGT CATACGGCAT CCCCAAGTCT 550
GATTGCGACG CTGCATTTGA CACCCTTTCC GATGGTGGCA AGACCATGGT 600
GACCCGAGAG ATTTTCGCCC GTCTCTGGAC AGAATATTTT GTATCTAACG 650
ACCGTGGGCT AAGGGCAACC ATCTGTTTGG TACCCTAAAA CTTTAAGAAA 700
AAACCTATCC CTTATTTTTA ATATATTCAT AAGAAATGTG AAAGAAAAAA 750
AAACTCTACC CCCAAAAAAA CTTCAACAAA GTTACTAACG TTTTTTCTAT 800
CAATTTTTCA AATTTCCTTT ATTATCGACC TATTGAGCTC TCTATCTTTT 850
ATCTATCCGA AGTCGGCACAA TATTTGCAAA ATTTAATTAA CTAATTGATT 900
AACAAAATAA CTAATTACGT TCAACGCTTA GTAATATGCA TAGCATGTGC 950
AGGATTTTTA ACCTTTTCTT TAAACTAATT ATTTAGTATA GCCATACGTT 1000
TTTTAAATAT ATATGTATGT ATATATATGT ATATATATAT ATATATGTGT 1050
GTATATATAT ATATATATAT AAATGGGACA TAGATTTTGT GAATATATCT 1100
TTGTCAATCC AATTGGACGG CAAGAACTGA TAGCCTTAAA TTTTCAATTT 1150
ACATCTTCAT ATAAACCTTT TCAATGAAAA CATATAAAAA ATAATATTAT 1200
AATTCATAAC TAATTCTTAT TAATATCATT AACTATAATA TACGTTGATT 1250
ATGAATTC
```

```
cp20    MAAHQLSDFQKNKILRVFNTSYDCNHDGVIEWDDFELAIKKICNLHSWPTD
scp1        GLNDFQKQKIKFTFDFFLDYNKDGSIQWEDFEEMIKRYKEVNKGSLS

cp20    KKHNEARATLK--LIWDGLRKYADENEDEQVTKEEWL-KMW-AECVKSVEK
scp1    ADYKSMQASLE--DEWRDLKGRADINKDDVVSWEEYL-AMW-EKTIATCKS
arf     ..ERISEAHEELQRMLNEDELRD-ALLLV--FANKQDLPNAMNAAEI-TDKLG
sar     ..ERFDEARVELDALFNIAELKD-VPFVI--LGNKIDAPNAVSEAEL-RSALG

cp20    ESLP--EW---LTKYMNFMFDVNDTSGDNIIDKHEY-STVYMSY-GIPKSD
scp1    ADLP--AW---CQNRIPFLFKGMDVSGDGIVDLEEF-QNYCKNF-QLQCAD
arf     HSLRHRQW------YIQATCA---TSGDGLYEGLEWLSTNLKNQ | ETVEY
sar     LNTTGSQRIEGQRPVEVFMCSVVMRNGY--LEAFQWLSQYI    | EELAI

cp20    DAAFDTLSDGGKTMVTQEIFARLWTEY--------FV--SNDRGLRATI-C
scp1    PAVYNVITDGGKVTFDLNRYKELYYRLLT---------SPAADAGNTLMG
arf     NISFTVWDVGG-----QDKIRPLWRHYFQNTQGIIFVVDSNDRE-RISEAH
sar     NIKFTTFDLGGHI---Q--ARRLWKDYFPEVNGIVFLVDAADPE-RFDEAR
```

```
Consensus    [A,G]XXXXGK[S,T]    NKXD    EXSAX
sar1p            G   LDNAGK    T      NKID    EELAI
cp20             T   LSDGGK    T      NIID    FELAI
```

Figure 22. *Sequence of squid cp20 cDNA (upper panel); the coding region is in boldface type. Aminoacid sequence of cp20 compared with* Amphioxus *scp1, yeast sar1p and arf1 (middle panel); aminoacids found by peptide sequencing of the tryptic digest are underlined. Possible GTP-binding sites in cp20, compared with homologous regions in sar1p (lower panel).*

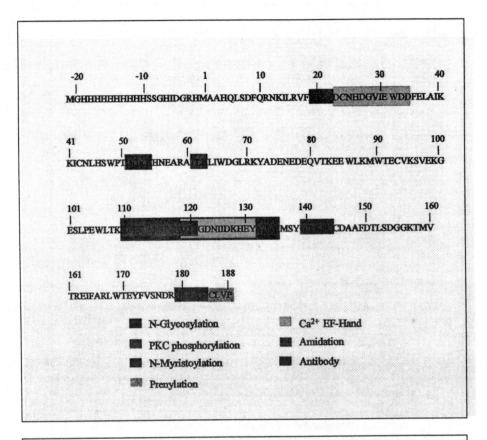

Figure 23. *Primary structure and possible post-translational modifications of cloned cp20. Tryptophan residues are marked in blue. Possible casein kinase consensus and GTP binding sites are not evidenced.*

The presence of additional basic residues at the N- or C-terminal of the target aminoacid (here, an arginine in position 59) is known to enhance the V_{max} and K_m of the phosphorylation reaction. Six casein kinase II phosphorylation sites (20-23, 82-85, 102-105, 120-123, 143-146, 151-154) are also predicted, but experiments did not confirm this hypothesis.

Two possible myristoylation sites are also present (139-144 and 179-184). An appreciable number of eukaryotic proteins are acylated by the covalent addition of myristoate (C14-saturated fatty acid) to their N-terminal residue via an amide linkage. The myristoyl moiety is often used to facilitate membrane binding (e.g., Di Luca, 1992, and references therein), and the reaction is catalyzed by the enzyme myristoyl-CoA:protein N-myristoyl transferase.

A polyisoprenylation consensus (CAAX box) domain is possible at the C-terminal (positions 185-188). CAAX is a motif shared among GTP-binding proteins used for attachment of either a farnesyl or a geranyl-geranyl tail, which also facilitates membrane binding. The modification occurs on cysteine residues three positions away from the C-terminal extremity. Since cp20 is known to translocate to the membrane under certain conditions (Nelson and Alkon, 1995), but no transmembrane sequences are found, it is reasonable to assume that either the CAAX box or the myristoylation sites are responsible for translocation.

Finally, three other potential post-translational modifications are present on the sequence, two asparagin-glycosylation (positions 19-22 and 118-121) and one amidation site (51-54).

Cp20's sequence also shows a type of non-covalent modification, two EF-hand calcium-binding domains (positions 23-35 and 119-131, fig. 23). Many calcium-binding proteins belong to the same evolutionary family (Nakayama et al., 1992), sharing a motif consisting of a 12 residue loop flanked on both sides by a 12 residue α-helical domain (for a review, Moncrief at al., 1990). In an EF-hand loop[2] the calcium ion is octahedrally coordinated. The six residues involved in the binding are in positions 1, 3, 5, 7, 9 and 12; these residues are denoted by X, Y, Z, -Y, -X and -Z (e.g. Travè et al., 1995). Positions X, Y and -Z are the most conserved (Zimmer et al., 1995). The relationship between EF-hand structural characteristics and biological activity is discussed in chapter 3.

The amino acidic composition of cp20 is summarized in table 2. Interestingly, a high number of tryptophan residues is present (fig. 23, W in blue, and tab. 2). These amino acids can influence the polarity and the steric hindrance of active sites, and seem to be widely distributed on the primary structure. Some other physical-chemical properties can be estimated from the composition of the 209 aminoacids, such as the instability index (37.30), the aliphatic index (65.31) and the extinction coefficient in denaturing conditions (49335 $M^{-1}cm^{-1}$ at 279 nm).

From the aminoacidic sequence it may be noticed that the old polyclonal antibody (chapter 1) was raised against a peptide matching cp20 in 9 out of 10 aminoacids. The low specificity could be due to this partial mismatch. A new polyclonal antibody was raised against a peptide corresponding to the sequence 109-135 (fig. 23). This antibody recognized cp20 as a 22 kD band in squid optic lobe extract and as a 25 kD band in engineered E. coli extract (fig. 24).

---

[2] The name EF-hand derives from the hand-like shaped positions of parvalbumin helices E and F, as discovered in the crystal structure by Kretsinger and coworkers in 1973 (for a hystorical overview, see Ikura, 1996).

Ala	(A)	10	4.8%		Leu	(L)	14	6.7%
Arg	(R)	9	4.3%		Lys	(K)	15	7.2%
Asn	(N)	10	4.8%		Met	(M)	7	3.3%
Asp	(D)	19	9.1%		Phe	(F)	8	3.8%
Cys	(C)	5	2.4%		Pro	(P)	4	1.9%
Gln	(Q)	3	1.4%		Ser	(S)	13	6.2%
Glu	(E)	15	7.2%		Thr	(T)	13	6.2%
Gly	(G)	12	5.7%		Trp	(W)	7	3.3%
His	(H)	17	8.1%		Tyr	(Y)	7	3.3%
Ile	(I)	14	5.3%		Val	(V)	10	4.8%

Table 2. *Composition of the 209 aminoacids of cloned cp20. There are 34 anionic residues (Asp+Glu) and 24 cationic ones (Arg+Lys).*

The same bands were detected for the native and the cloned purified proteins. No bands were visible in rabbit cortex[3], and both controls (prebleed and preabsorbed) showed a blank response, confirming the specificity of the signal.

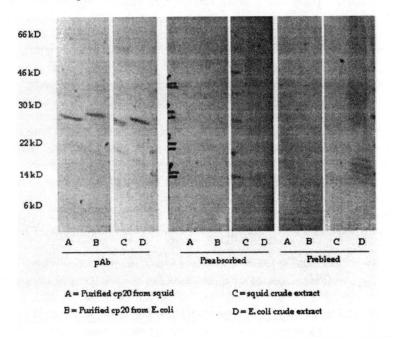

A = Purified cp20 from squid   C = squid crude extract

B = Purified cp20 from E.coli   D = E.coli crude extract

Figure 24. *Characterization of the pAb. MW are marked in the middle lanes.*

---

[3] The antigenic sequence lies on the possible consensus site for N-glycosilation, a post-translational modification which is common in vertebrate species neurons. If mammalian cp20 is glycosilated by a large group (such as syalic acid), it would not be recognised by the antibody independently of the aminoacidic homology.

The washing and developing system described in the Materials and Methods was optimized to achieve maximum antibody specificity.

A large scale purification of cloned cp20 was set up. *Escherichia coli* cells were grown in standard conditions (fig. 25) and the expression of cp20 was induced by the addition of IPTG. This non-hydrolizable analogue of lactose triggered the induction of the cp20 gene, which was mounted on the commercial *lac* vector (Nelson et al., 1996b). The resistance to ampicillin of the cell line was also a result of genetically engineered E. coli; non-expressing cells were killed by the addition of the antibiotic to the medium. The use of single colonies and the intermediate step of growing a mother solution also increased the specificity.

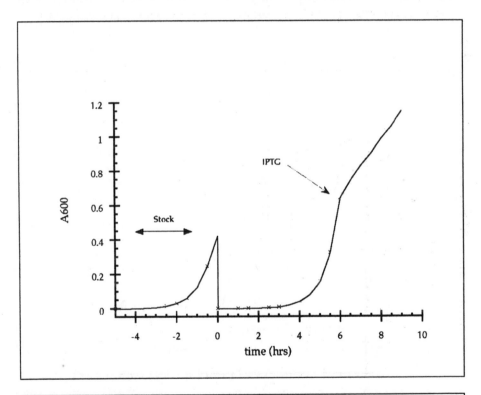

Figure 25. *Growth kinetics of ampicillin-resistant (and cp20-expressing) E. coli. If the cellular growth had a different shape, the preparation was discarded for the possible presence of other contaminating bacterial strands.*

The time constant τ_2 of cell growth was 30 minutes at 37°C, and the addition of IPTG remarkably reduced the proliferation rate, thus confirming that the expression was effective. Western blot experiments showed that non-engineered E. coli cells did not express cp20, while non IPTG-induced cells expressed a reduced amount of the protein.

The crude extract was prepared according to standard procedures (see experimental part), and the addition of protease inhibitor cocktail was necessary to avoid proteolytic degradation. The crude extract was fractionated on a commercial his-tag low pressure affinity resin, precharged with Ni^{2+}. The resulting chromatogram (fig. 26) shows a high-affinity retained peak, which is missing in preparations carried out using non-engineered cells or non-IPTG stimulated cells, under the same conditions. The chromatographic shape and time pattern were reproducible both between successive loads in the purification and between different preparations.

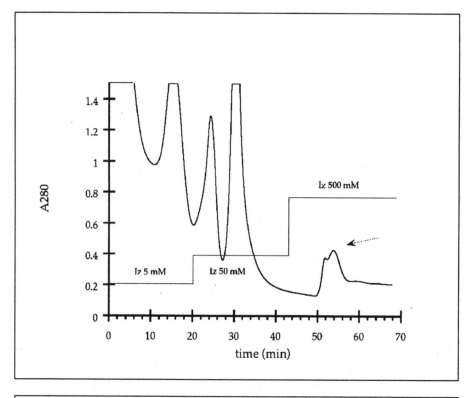

Figure 26. *His-tag chromatogram of E. coli crude extract supernatant. The arrow below the second imidazole jump indicates the cp20-containing peak.*

Peaks eluting at low or medium imidazole concentrations were tested by western blot, and contained a relatively low quantity of cp20. The high-imidazole peak was extensively desalted to allow the next metal-affinity chromatographic step. Some precipitation occurred upon concentration. The same effect had been observed in the native squid cp20 purification (chapter 1), suggesting the occurrence of aggregation phenomena. The precipitated proteins were soluble in 1% SDS or 8 M guanidine hydrochloride, and showed several bands on silver stained SDS-PAGE analysis, with a major component at 25 kD.

The supernatant was loaded on IMAC-HPLC. The column reproducibility was tested with myoglobin; the high speed of perfusion chromatography allowed the processing of 4-5 cp20 samples per preparation (fig. 27).

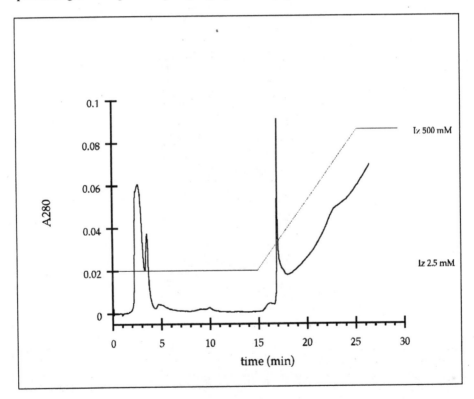

Figures 27. *IMAC-HPLC elution of partially purified cloned cp20. The signal drifts follows the imidazole gradient despite the use of an "optically pure" reagent.*

Fractions were collected every 0.5 minutes and analyzed by dot-blot and immunostaining with the polyclonal antibody (fig. 28). The resulting cp20-

containing peak was distributed over the imidazole gradient with a peak at about 160 mM Iz.

Cp20-containing fractions were pooled together, desalted, concentrated and stored in small aliquots for a prolonged time. An SDS-PAGE analysis revealed an excellent purity of the protein.

The silver stained gels and the western blots from the fractions of each purification step are reported in fig. 29.

A summary of the yields and recoveries of each step of the preparation of cloned cp20 is shown in table 3.

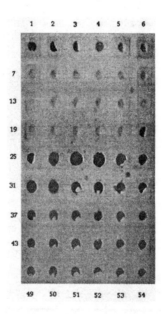

Figure 28. *Immunostained dot-blot analysis of the IMAC chromatogram. Fraction numbers correspond to 0.5 minute units on fig. 27.*

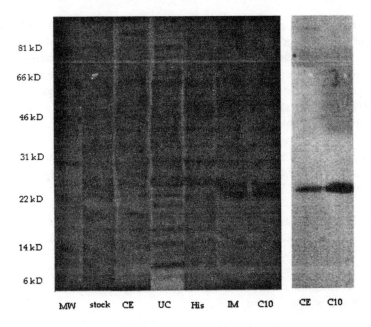

Figure 29. *SDS-PAGE of the fractions after each step of the purification. Silver stain, left panel; western blot, right panel. Stock, E. coli cells before IPTG induction; CE, crude extract; UC, ultracentrifuged; His, histidine-tag retained fraction; IM, IMAC-HPLC retained fraction. C10, final, pure, desalted sample.*

	TotProt	TotCp20	Purity %	StepRec	StepPF	TotRec	TotPF
CE	2050000	4100	0,2	1	1	1	1
UC	450000	3600	0,8	0,88	3,5	0,88	3,5
His	2955	1980	67	0,55	46,1	0,48	161,8
[PP]	[690]	[449]	[63,8]	[0,23]	[0,22]	[0,11]	[34,9]
IM	595	505	85	0,26	0,32	0,12	52,3
C10	490	480	98	0,95	1,1	0,12	57,4

Table 3. *Purification data. Vertical titles (samples): CE, crude extract; UC, ultracentrifuged; His, his-tag retained; PP, precipitated during concentration after his-tag; IM, IMAC-retained; C10, final, desalted, pure sample. Horizontal titles (parameters) are defined in table 1. TotCp20 (in μg) was estimated by silver stained SDS-PAGE and immunostained western blot. StepRec and StepPF for PP and IM are referred to the corresponding values of His.*

The efficiency of this purification is impressive when compared to the isolation of the natural protein presented in chapter 1 (tab. 1), as expected. The purity is already fairly high after the his-tag step, and indeed several electrophysiological experiments were carried out on that crude fraction. For other experiments, a second passage on the same his-tag column was performed (Nelson et al., 1996b). The HPLC-IMAC step was necessary though to conduct a spectroscopic characterization (chapter 3), which requires homogenous samples.

The quality of the expression vector is a limiting factor for the yield of the purification. The same bacterial system could yield over 10 mg of protein after optimization (see e.g. Warren et al., 1994; Weber et al., 1994).

An AX-300 HPLC step was also tested on the his-tag retained fraction. Several samples were collected during the potassium acetate gradient, and loaded on a gel (fig. 30). Cp20 did not appear as a band at 25 kD in any of the tested fractions. The column was extensively washed with 1M buffer, and the resulting sample was desalted, concentrated and loaded on the gel (last lane of fig. 30), but the protein did not elute from the column.

It is possible that the fused histidine tail drastically increases the affinity of cp20 for the anion-exchange resin; in these conditions, the recovery is difficult, and this chromatographic passage was not employed further in the purification or analysis of the cloned protein.

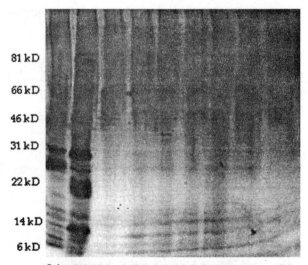

Figure 30. *Silver stained SDS-PAGE of AX-300 HPLC fractions from the elution of cloned cp20. Concentrations are referred to Kac at the column output.*

Cp20 shows two potential EF-hand calcium binding sites in the aminoacidic sequence, as discussed at the beginning of this chapter. The purified protein was tested for calcium binding activity, by means of overlay blot (fig. 31).

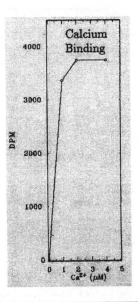

Figure 31. Ca^{2+} *binding as a function of cp20 concentration (Nelson et al., 1996b).*

The experiment clearly indicated a stoichiometry of 2:1, i.e. both calcium binding sites are active. Binding was performed in the presence of Mg^{2+} and the absence of GTP. Numerous other control proteins did not bind Ca^{2+} under these conditions. The specificity of the effect was also confirmed my the molecular weight on the nitrocellulose membrane (25 kD), although some degradation product was visible (Nelson et al., 1996b).

It was also possible to test calcium binding under physiological conditions, and to filter the solution through a nitrocellulose sheet[4]. The scintillation counting of membrane bound ^{45}Ca versus passed through ^{45}Ca could be analyzed by means of classical Scatchard methods (see e.g. Ascoli, 1993, and references therein), which

---

[4] Proteins have a high affinity for nitrocellulose, while the other solution's components pass through. Calcium is partitioned between bound fraction (with the protein, on the membrane) and free fraction (with the solution, passed through).

confirmed a two-site interaction and indicated a dissociation constant of ≈40 nM for the higher affinity site (fig. 32).

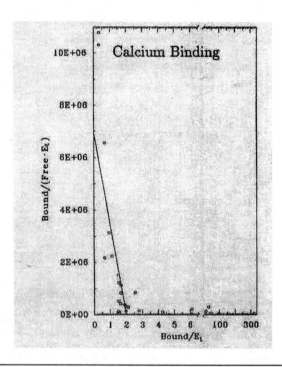

Figure 32. *Scatchard analysis of cp20 calcium binding (Nelson et al., 1996b), suggesting the presence of one low-affinity (Kd ≈ 400 nM) and one high-affinity site (an order of magnitude stronger).*

This result is in good agreement with the properties of scp-like proteins (e.g. Findlay and Sykes, 1993, and references therein).

Cloned cp20 was an unusually high-affinity substrate for α-PKC with a Km of 5.4 nM (Nelson et al., 1996b), similar to the Km of 7.1 nM reported for native cp20 (Nelson and Alkon, 1995). Addition of 1mM GTP, a concentration that only slightly inhibited phosphorylation of other substrates (e.g. less than 30% inhibition of histone phosphorylation; Nelson et al., 1996b), completely inhibited phosphorylation of cp20 (fig. 33).

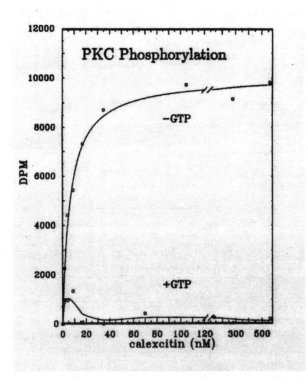

Figure 33. *Phosphorylation of cloned squid cp20 by α-PKC in the absence (upper curve) and presence (lower curve) of GTP (Nelson et al., 1996b).*

Microinjection of cloned cp20 in Hermissenda type B photoreceptor neurons caused a marked increase in excitability which was manifested as a long-lasting depolarization (fig. 34). A similar depolarization was previously observed after injection of comparable amounts of purified native squid cp20, and after associative conditioning (see parts 2 and 3 of the Introduction). The normal response to a 1 second flash of light returns to the original resting potential (*top* of fig. 34). Injection of cp20 depolarized the cell for more than 5 minutes (*bottom*, only one minute is represented). Injection of heat-inactivated cp20 did not alter the light response.

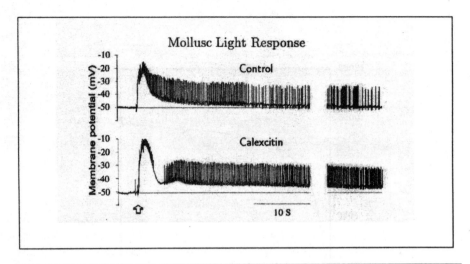

Figure 34. *Intracellular recordings from Hermissenda B-photoreceptor in response to a flash of light before (top) and after (bottom) injection of purified cloned cp20. Time scale at right is compressed 3x (from Nelson et al., 1996b).*

Previous results have established that the effects of cp20 and conditioning are both mediated by inhibition of the outward K^+ current in i_A and i_{Ca-K} channels (Alkon, 1994; see also the Introduction). Cp20 microinjection in rabbit Purkinjie cells also produced a substantial increase in membrane excitability (fig. 35).

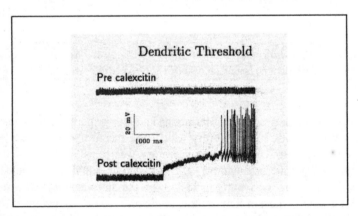

Figure 35. *Membrane excitability increase due to cp20 (Nelson et al., 1996b). Passively propagated somatic spikes (small amplitude) but not dendritic spikes (large amplitude) occurred spontaneously in cerebellar Purkinjie cells before the addition of cp20 (upper). Local dendritic Ca^{2+} spikes occurred spontaneously after injection of cp20 through the recording electrode (lower trace).*

First, there was a 25% reduction in the current required to elicit local, dendritic calcium spikes following injection of cp20, compared to a 3.9% increase in that current following control injections (Nelson et al., 1996b). Second, there was a noticeable increase in spontaneous dendritic spikes following injection of cp20. In contrast, cp20 had no effect on membrane potential, input resistance, or current required to hyperpolarize the membrane 20 mV below somatic spiking. The role of rabbit cerebellum in associative memory and synaptic plasticity mechanisms could also be mediated by modulation of calcium-dependent potassium channels, and therefore by cp20 (Schreus et al., 1991).

The electrophysiological effect of cp20 was also tested on unitary K^+ channel conductances in inside-out patches excised from cultured human skin fibroblasts. Bath addition of cp20 reduced the mean channel open time and the mean open probability (fig. 36). The current had a reversal potential consistent with K^+ conductance (Nelson et al., 1996b), and had unitary conductances (110-130 pS) that fall within the range of the previously characterized channel which is absent in cells from patients with Alzheimer's disease (Etcheberrigaray et al., 1994).

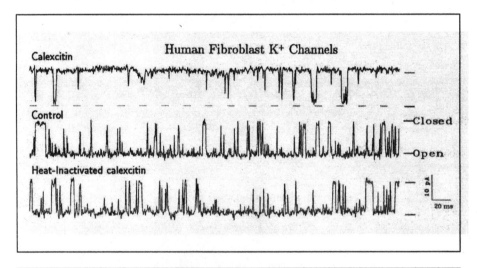

Figure 36. *Single-channel current traces of K^+ channel activity in an excised inside-out membrane patch of human skin cultured fibroblast recorded in the absence of cp20 (middle trace), after addition of heat-inactivated cp20 (lower trace), and after addition of intact cp20 (upper trace). Cp20 markedly reduced the mean open time and probability of openings (Nelson et al., 1996b).*

No change in potassium unitary currents was observed for over 10 minutes after addition of heat-inactivated cp20.

Since calcium-dependent potassium channels are conserved among species, it is likely that the mechanism of action of cp20 is similar in Hermissenda, squid, rabbit and human K^+ currents. The strong inhibitory effect showed by these results also suggests that a mammalian analogue of cp20 may exist in rabbit and human CNSs, and could play a role in the normal regulation of neural channels.

Immunocytochemical results in squid optic lobe demonstrated that cp20 was localized predominantly in the plexiform layer, which contains fibers originating from the retina and optic nerve, as well as amacrine cells with spreading tangential processes (Nelson et al., 1996b). A small percentage of neuronal cell bodies also stained intensely for cp20. This localization in regions of terminating axonal processes is consistent with previously observed effects of cp20 on axonal transport and neuronal branching (Moshiach et al., 1993). In order to probe the identity of squid optic lobe cp20, cloned cp20 and axon-localized cp20, a western blot analysis was performed on the purified proteins and tissue crude extracts (fig. 37). Natural squid and axon proteins were detected as comigrating bands (22 kD), while cloned E. coli protein showed a single band at 25 kD, as expected.

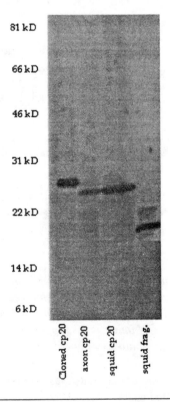

Figure 37. *Western blot of purified bacterial cp20, squid axon crude extract, squid optic lobe crude extract and purified fragment of squid cp20 (chapter 1).*

In order to further test that these proteins belonged to the same structural family, a reverse-phase chromatographic experiment was set up. Protein crude extract from squid axon and purified cloned cp20 were loaded; the partially purified squid optic lobe 22 kD intact protein and the 18 kD fragment (chapter 1) were also injected. Fractions were collected, lyophilized and analyzed by SDS-PAGE and western blot (fig. 38).

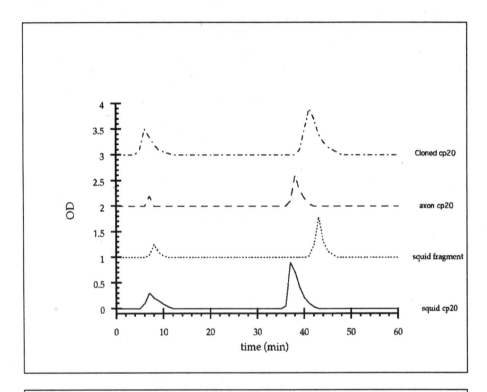

Figure 38. *Reverse-phase HPLC elution and dot-blot immunodetection of purified cloned cp20 (centered trace, baseline at 3.0), squid axon crude extract (broken trace, baseline at 2.0), squid optic lobe purified 18 kD fragment (dotted trace, baseline at 1.0) and purified intact squid optic lobe cp20 (full trace, baseline at 0.0). Optical density of blot scan is in arbitrary units.*

The similarity of retention times suggests that these proteins have an equal polarity, indicating that they are related or identical molecular entities. An experiment of the same kind showed that also using rat liver crude extract as a starting tissue, an immunoactive peak showing a band at 20 kD elutes at the same retention time as invertebrate cp20. Experiments are in progress to sequence the corresponding mammalian protein.

Calexcitin/cp20 is the only protein discovered so far that binds both calcium and GTP. The crucial cellular role that the Ca^{2+} and GTP biochemical pathways play in neurons make cp20 a unique candidate for the role of convergence point which is necessary for associative memory mechanisms (Alkon, 1989, 1993; Kandel, 1993; Dudai, 1989). What the precise effect of GTP- and calcium-binding on cp20 activity is has not been discovered yet, nor is it known how the PKC pathway is involved. A complete study of cp20's structure is needed to begin answering these questions. In the next chapter we describe the characterization of cp20 secondary structure and its modulation by calcium binding.

CHAPTER 3

Conformational study of cp20

The fact that cp20 is chemically modified upon associative conditioning led to its initial discovery. The PKC-induced phosphorylation of cp20 causes its translocation into the cellular membrane, where cp20 can modify the neuronal currents. Several questions need to be answered. Is cp20's effect on channels direct or mediated by other molecules? Is cp20 translocation itself mediated by other molecules? Are there other signals that modulate cp20's activity? If cp20 acts as a cellular switch (as the causal relation between its phosphorylation status and cellular activation suggests) are there two relatively stable conformations corresponding to an "active" and an "inactive" state?

The first step towards a complete understanding of the cellular action mechanism of cp20 was its primary structure characterization. It is not yet known which possible post-translational modifications of potential signaling interest (such as the myristoylations) occur *in vivo* or *in vitro*, but the experimental data on phosphorylation, GTPase activity and calcium binding nicely match the theoretical expectations.

In this chapter, we describe the characterization of cp20's secondary structure. Circular dichroism, Fourier transform infrared spectroscopy and sequence-matching computational algorithms yielded a set of conformational domains in good agreement with each other. CD was particularly suitable to study the effect of calcium on the secondary structure. Ca^{2+} binding caused a marked variation of cp20 conformation; this variation may be interpreted as a decrease of the α-helical content and a parallel increase of β-sheets. This one-phase transition, which seems to occur in a physiologically meaningful range of calcium concentrations, was also independently confirmed by fluorescence spectroscopy and non-denaturing gel electrophoresis. All these experiments indicated that cp20 undergoes a partial unfolding when calcium is removed. The overall protein polarity is not affected by the conformational transition.

Cloned cp20 was also chemically modified by enizmatically removing the histidine tail; the product could be purified by a second passage through IMAC-HPLC, and showed a molecular weight identical to that of the squid optic lobe natural protein. Calcium binding properties and secondary structure were not affected by the histidine tail removal.

The effect of the PKC-induced phosphorylation of cp20 was also investigated, but the low quality of the commercial enzyme and the low residual amount of protein only allowed a preliminary characterization.

Finally, the study of cp20's conformational stability was performed. At room temperature, the secondary structure of cp20 slowly but consistently changed with an increase of the random structure and the β-sheet and a reduction of the α-helix.

A study based on solid-state Fourier transform infrared spectroscopy also confirmed the occurrence of a concentration-dependent aggregation process, which might explain the precipitation phenomena observed during the purifications of both natural and cloned protein.

Materials and Methods

Mass spectra were recorded on a Perkin Elmer API Sciex Ion-spray mass spectrometer. Cation formation was promoted by diluting the protein solution in a 10% methanol solution containing 0.1% HCl, and analyzed by direct infusion or by on-line RP-HPLC.

UV spectra for protein quantitation were recorded on a Varian Cary 4E double-ray spectrophotometer at a scan speed of 0.1 nm/min, using storing buffer (5 mM trisCl pH 7.5, 25 mM NaCl, 5 mM imidazole, 0.1x protease inhibitor cocktail solution) as a blank.

Vacuum UV circular dichroism spectra were recorded over the range 178-260 nm using a Jasco J700 (Jasco Instruments, Tokyo, Japan) spectropolarimeter. A Jasco J720 and a Jasco J600 instruments were also used for some experiments. Measurement were made using a 0.01 cm quartz disassemblable cell (V=20 µl) or a 0.02 cm quartz jacked cell (V=15 µl). Cells were rinsed with water extensively before each measurement, and solutions were recovered by either disassembling the cell or using an electrophysiology needle. Spectra were recorded in triplicate at $14\pm1°C$ (Shimadzu water bath), after equilibrating the instrument cell chamber with N2 at a flow rate of 15 l/min. The time constant was 3 s, the scan rate 10 nm/min and spectral slit width 1 nm. All blanks were recorded in the same conditions and subtracted from the sample spectrum.

Spectra were cut at 180 or 182 nm before deconvolution, and analyzed by single value decomposition with Varsel software (Hennessey et al., 1987) running on a Unix machine. The complete basis of 33 proteins was used in 528 subgroups of 31 elements each, and the search was optimized using protein concentration as a variable parameter.

Solutions were prepared by at 0°C diluting 5 µl of freshly thawed protein stock solution (C10 fraction from chapter 2) with 5 µl of protease inhibitor cocktail and 5 ml of calcium buffer (final protein concentration: about 0.15 g/l). Calcium buffer was prepared using the EGTA software written by T. J. Nelson and available by anonymous ftp in binary mode at las1.ninds.nih.gov in the pub/dos directory. Blank solutions were obtained by substituting the protein stock solution with storing buffer.

Computation for secondary structure prediction was performed using a multiple sequence alignment algorithm (Sopma), available on the internet (for details, deleage@ibcp.fr), using default parameters and high accuracy (evaluated as 72%). A simple query of percentage of alpha, beta and random structure was submitted to maximize reliability (Geourjon and Deleage, 1994 and 1995).

Fluorescence measurements were obtained in triplicate at room temperature using a SLM-Aminco Model 500 fluorescence spectrophotometer (Aminco, Corvallis, Oregon, USA). Protein solutions were recovered from CD analysis, diluted with protease inhibitors and buffered with calcium to 3 ml in 1 cm quartz cells (final protein concentration: 0.03 g/l). The excitation wavelength was 279 nm, the scan range 300-400 nm, the speed of 10 nm/min. The instrument was calibrated with 1 mM fluorescein, following the manufacturer's protocol.

Phenyl-sepharose chromatography was performed at room temperature using 2 ml of fast flow resin (Sigma, St. Louis, MO, USA) packed in a 5 ml polypropylene column (Pharmacia, Uppsala, Sweden), at a constant flow rate of 0.5 ml/min, loading 0.1 ml of sample and collecting 1 ml fractions for dot blot analysis.

Native polyacrylamide gel electrophoresis was performed using a precasted 4-20% gel with a Novex apparatus (chapter 1) in non-denaturing conditions (sample buffer: 1.5 M trisCl pH 8.8, 4 ml; 0.1% bromophenol blue, 0.5 ml; glycerol, 2 ml; H_2O to 10 ml. Running buffer 10x: tris base, 29 g; glycine, 144 g; H_2O to 1 l). Gels and western blot were run and stained as described in previous chapters.

The oligo-histidine tail was removed from cp20 by a catalytic reaction promoted by Bovine ImmunoPure Factor Xa (Pierce, Rockford, IL, USA). 50 μg of cp20 in 180 μl of water were mixed with 5 μg of Factor Xa in 3 μl of buffer (5 mM trisCl, pH 6.0; 0.5 M NaCl; 1mM benzamidine hydrochloride). No protease inhibitors were added to avoid protease inhibition. The reaction mixture was incubated at room temperature for 2 hours and an aliquot analyzed by SDS-PAGE (silver stained and western blotted). The reaction was stopped by dilution with protease inhibitor cocktail and two passages through a Centricon C50, which eliminated Factor Xa (MW = 47000). The resulting mixture was repurified by IMAC-HPLC (as described in chapter 2): the void volumes and the retained fraction, containing reacted (without histidine tail) and unreacted (with histidine tail) cp20, respectively, were saved, concentrated, desalted and stored at -80°C after SDS-PAGE analysis.

Recombinant α-PKC from rabbit brain expressed in baculovirus system was purchased from Upstate (Biotechnology Inc., Lake Placid, NY, USA), and used without further purification. For phosphorylation experiments, 40 μl of reaction buffer (20 mM trisCl, pH 7.2; 25 mM β-glycerophosphate; 2 mM EDTA; 10 mM EGTA: 10% glycerol; 10 mM β-mercaptoethanol; 1mM DTT; 5 mM benzamidine; 1x protease inhibitor cocktail) were mixed with 20 µl of lipid activator (0.5 g/l phosphatidylserine and 0.5 g/l diacylglycerol in chloroform) previously air-dried. 5 mM (final) $CaCl_2$, 2 µl of dimethylsufoxide, 10 mM (final) $MgCl_2$ and 1x phosphatase inhibitor cocktail (2 mM sodium orthovanadate; 200 mM NaF) were added, and the suspension was sonicated in ice for one minute. The solution was divided in 10 µl portions (to test several samples) and 50-200 ng of cp20 were added. Finally, 25 ng of PKC and 0.5 µl of ^{32}P-ATP (3000 Ci/mMole, three month-old) or 0.5 µl of 10 mM cold ATP plus 12 ng of PKC (0.5 µl) were added to each tube. Radioactive samples were incubated for variable lengths of time at 31°C and spotted on phosphocellulose paper. The sheets were washed twice for 30 minutes in 250 ml of 75 mM H_3PO_4 and transferred into a vial containing scintillation cocktail for ^{32}P counting. Cold samples were analyzed by SDS-PAGE, repurified on IMAC-HPLC, and the retained fraction was desalted, concentrated and stored at -80°C after SDS-PAGE analysis.

Fourier transform infrared spectra (FT-IR) were recorded on a Perkin Elmer 1760 Spectrophotometer equipped with a Perkin Elmer TGS detector and a Perkin Elmer M7300 computer. The instrument was continuously purged with dry air for 15 minutes before data collection and during measurement to eliminate vapor absorption. A shuttle spectrum was used to allow the background spectrum to be signal averaged over the same time period as the sample spectrum. For each sample, 128 interferograms were recorded, averaged, and Fourier transformed to yield a nominal resolution of 4 cm^{-1}. Aqueous solution analyses were performed by placing 3 µl of sample (with a protein concentration > 5% w/w obtained by air flow concentration) between two BaF_2 disks using a 9.3 µm thick Teflon spacer. Subtraction of water spectra (recorded in the same instrumental conditions) was judged to be appropriate when it yielded an approximately flat baseline (without negative side lobes) in the 1900-1720 cm^{-1} region and near 2130 cm^{-1}. Solid state analyses were recorded on air-dried samples on the BaF_2 disk. Slow evaporation was obtained by placing the sample under a glass bell overnight, while fast evaporation was achieved in a dryerite filled container.

Second derivative deconvolution procedures and data analysis were processed as described (Bramanti and Benedetti, 1996), with a Fortran Microsoft version 5.1 program, based on the VA08A/AD subroutine of the Harwell library, which implements the CGA (conjugated gradient algorithm) software.

Results and Discussion

A mass spectral analysis of pure cp20 was performed in order to assess the molecular purity of the sample, and to obtain information on the post-translational modification status of the protein. The ion-spray (IS) spectrum gave a single convergent peak series (fig. 39), converging to a molecular weight of 25338±1 D, in good agreement with SDS-PAGE data and calculated MW.

The ions producing the molecular weight deconvolution consisted of four consecutively charged molecules (21-24), with the highest one corresponding to the theoretical maximum of positive charges for cp20. All the values are within the estimation tolerance, thus confirming the reliability of the result. The intensity of the signal was nonetheless lower than the what had been expected for this experiment. Increasing the concentration of the protein in the sprayed solution dramatically reduced the signal, and dilution did not yield any benefit. Under these conditions, it was not possible to assess the post-translational status of cp20, although the result indicate that cp20 was the only protein in the solution. Matrix-assisted laser desorption ionization mass spectrometry (MALDI-MS) gave similar results, confirming the molecular weight and also indicating the presence of cp20 dimers in vapor phase (A. Raffaelli, Univ. of Pisa, Italy, personal communication), in agreement with a previously reported property of cp20 (Nelson et al., 1994).

In order to study cp20 secondary structure by CD and FT-IR, an accurate quantitation of the solution protein content was necessary. UV spectroscopy was used to measure cp20 concentration (Viegi, 1995), using the spectroscopical parameters (ϵ vs λ) calculated on the basis of the primary structure (chapter 2). The experimental UV spectrum of a native cp20 solution diluted in water (after blank subtraction) is reported in fig. 40.

The protein content was also assessed by the bincichonic acid (BCA) reaction (Smith et al., 1985). Two calibration curves were made in triplicate with bovine serum albumin and hemoglobin (fig. 41), and cp20 concentration was measured in the linear range. The calculated concentration of stock solution (0.175 g/l) was in excellent agreement with the UV experiment.

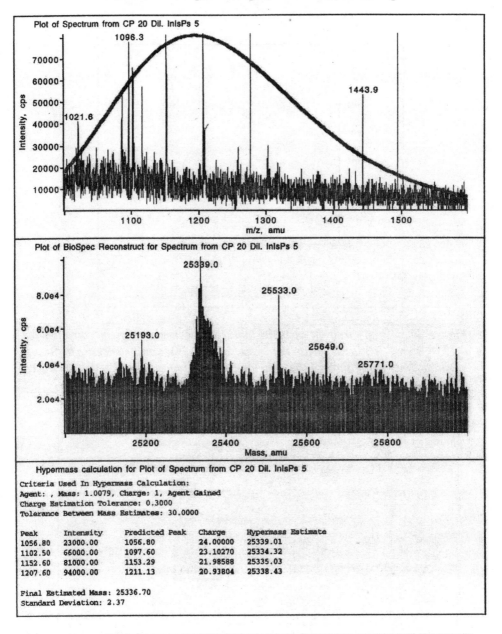

Plot of Spectrum from CP 20 Dil. InlsPs 5

Plot of BioSpec Reconstruct for Spectrum from CP 20 Dil. InlsPs 5

Hypermass calculation for Plot of Spectrum from CP 20 Dil. InlsPs 5

Criteria Used In Hypermass Calculation:
Agent: , Mass: 1.0079, Charge: 1, Agent Gained
Charge Estimation Tolerance: 0.3000
Tolerance Between Mass Estimates: 30.0000

Peak	Intensity	Predicted Peak	Charge	Hypermass Estimate
1056.80	23000.00	1056.80	24.00000	25339.01
1102.50	66000.00	1097.60	23.10270	25334.32
1152.60	81000.00	1153.29	21.98588	25335.03
1207.60	94000.00	1211.13	20.93804	25338.43

Final Estimated Mass: 25336.70
Standard Deviation: 2.37

Figure 39. *IS-MS spectrum of cp20 (upper panel). The deconvoluted spectrum (middle panel) is based on a single convergent series of peaks (lower panel).*

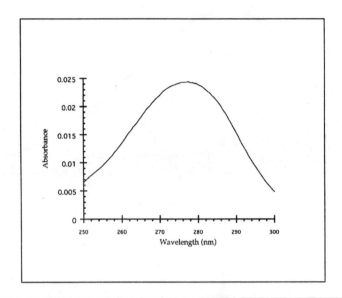

Figure 40. *UV spectrum of cp20 in the aromatic region. A concentration of 0.175 g/l can be estimated for the stock solution from an extension coefficent calculated assuming that all the cysteines are in reduced (ProtParam software, chapter 2).*

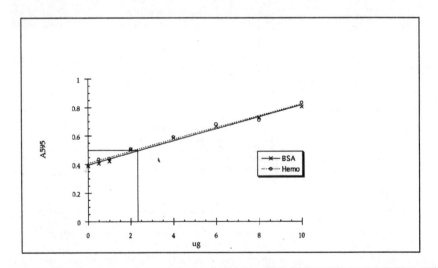

Figure 41. *BCA protein test. BSA and hemoglobin behave similarly. Cp20 concentration is indicated by perpendicular lines (A=0.498, c=2.65 µg).*

A CD spectrum of cp20 solution was recorded at a 1:4 to 1:10 dilution of the stock solution, and concentration did not affect spectral shape within this range. CD spectra were deconvoluted according to Johnson and coworkers (see Materials and Methods), and the secondary structure evaluated in five contributions (α-helix, antiparallel β-sheet, parallel β-sheet, turns and random coil, including "other structures"). The spectrum corrected for baseline and the deconvoluted secondary structure are reported in fig. 42.

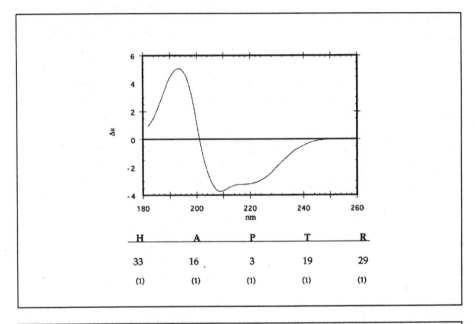

Figure 42. *CD spectrum (reported as molar ellipticity) of cp20 in the 180-260 nm range and deconvoluted values of secondary structure. H, α-helix; A, antiparallel β-sheet; P, parallel β-sheet; T, turns; R, random coil and other structures.*

All secondary structure component values in fig. 42 are optimized by the SVD algorithm (see § 5 of chapter 1 and *Materials and Methods* of this chapter) and have a tolerance of ± 1% (average of three measurements).

The secondary structure of cp20 was quite unstable, as judged by the rapid change of the CD spectrum over the course of a few hours at room temperature. In particular, CD spectra were recorded on solutions left at room temperature for a variable time were deconvoluted, and the secondary structures corresponding to several of these samples are reported in tab. 4.

Time (hrs)	H	A	P	T	O
0,5	33	15	3	19	30
1	30	14	3	12	41
2	28	12	2	13	45
4	25	10	1	8	56
8	24	9	1	4	62
16	16	7	0	9	68

Table 4. *Variation of cp20 structure in a solution at room temperature. A CD spectrum is recorded in about 30 minutes (see Materials and Methods). Spectra recorded after 24 hours were not analyzable by single value decomposition.*

The above effect is likely to be the sum of a purely transitional phenomenon, i.e. only regarding the protein's conformation, and a proteolytic degradation effect. Several low-molecular weight bands appear in the silver stained SDS-PAGE if the purified protein from either squid optic lobes or E. coli is left for hours at room temperature. In order to reduce this effect, CD spectra were recorded at 15°C upon immediate thawing of solutions. A lower temperature caused water condensation on the cell walls even at high flow rate of N_2 in the instrument.

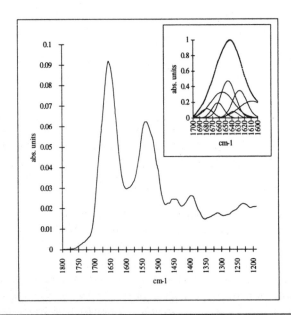

Figure 43. *FT-IR spectrum of cp20 solution (6 % w/w) in the 1800-1190 cm^{-1} region. Inset: results of the deconvolution procedure applied to the amide I region. α-helix: 30 %; β-sheet, 20 %; random/extended coils: 27 %; turns: 23 %.*

An FT-IR spectrum of cp20 solution was also measured (fig. 43) and the Amide I region was deconvoluted as described in the experimental part. The resulting contents of α-helix, β-sheet, random/extended coils and turns were in excellent agreement with the CD data.

Solid state FT-IR spectra were also recorded, by allowing a small amount of solution to dry on BaF_2. While fast evaporation gave results in good agreement with the solution data, slow evaporation yielded spectra with a shape dependent on the concentration of the starting solution (fig. 44).

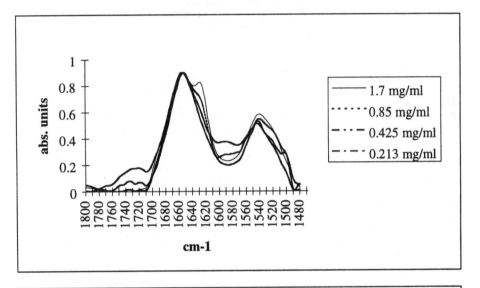

Figure 44. *FT-IR absorption spectra in the 1800-1460 cm^{-1} region of dried films obtained by spreading the solution of cp20 at different protein concentrations on BaF$_2$ windows and allowing them to evaporate overnight.*

This interesting effect might indicate a phenomenon of aggregation, which is also suggested by the precipitation of cp20 during the preparations from squid and E. coli.

By applying the deconvolution procedure to the Amide I region of the above FT-IR spectra, the intensity and frequency of each band component were obtained (fig. 45). The reliability of the method can be appreciated by observing that experimental and reconstructed spectra are almost indistinguishable for all the samples.

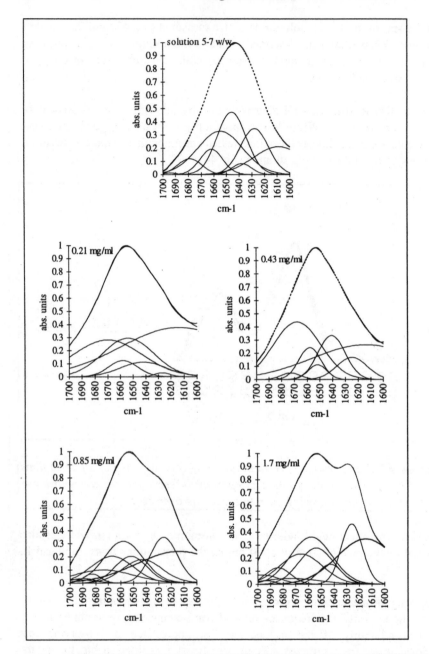

Figure 45. *Normalized spectra of cp20 (thin lines) compared to the spectra reconstructed by the algorithmic deconvolution to produce the best fit (thick lines). The Gaussian components are also shown. Upper panel, cp20 solution (from fig. 43). Other panels, films from solutions at the specified concentrations.*

Sample description	j	pj1*	%OD	Pj2*	xj*	Assignment
Solution (5-7% w/w)	1	0.01	1	1694	11	turns
	2	0.12	7	1679	22	turns
	3	0.19	11	1662	21	extended α-helix
	4	0.33	19	1655	42	α-helix
	5	0.47	27	1646	27	random structure
	6	0.08	5	1638	18	turns
	7	0.35	20	1628	26	β sheet
	8	0.21	12	1609	47	turns
Solid state (1.7 mg/ml)	1	0.07	3	1697	19	turns
	2	0.03	1	1693	15	turns
	3	0.13	6	1684	23	turns
	4	0.04	2	1674	53	turns
	5	0.23	11	1666	39	extended coil
	6	0.36	17	1658	38	extended α-helix
	7	0.28	13	1654	32	α-helix
	8	0.15	7	1643	22	random structure
	9	0.46	22	1626	20	β sheet
	10	0.35	17'	1615	54	turns
Solid state (0.85 mg/ml)	1	0.03	2	1698	34	turns
	2	0.03	2	1690	22	turns
	3	0.07	4	1683	16	turns
	4	0.10	6	1674	55	turns
	5	0.20	12	1667	39	extended coil
	6	0.32	19	1658	36	extended α-helix
	7	0.20	12	1652	30	α-helix
	8	0.18	10	1643	36	random structure
	9	0.35	20	1626	25	β sheet
	10	0.24	14	1614	79	turns
Solid state (0.43 mg/ml)	1	0.02	1	1685	5	turns
	2	0.05	3	1673	19	turns
	3	0.44	27	1668	43	extended coil
	4	0.24	15	1658	20	extended α-helix
	5	0.11	7	1652	17	α-helix
	6	0.34	21	1641	25	random structure
	7	0.17	10	1625	28	β sheet
	8	0.26	16	1614	101	turns
Solid state (0.21 mg/ml)	1	0.01	1	1689	12	turns
	2	0.01	1	1684	13	turns
	3	0.10	7	1673	17	turns
	4	0.39	28	1667	64	extended coil
	5	0.21	15	1658	21	extended α-helix
	6	0.05	4	1654	28	α-helix
	7	0.24	17	1642	34	random structure
	8	0.02	1	1627	17	β sheet
	9	0.38	27	1614	98	turns

Table 5. *j, Gaussian components; pj1*, peak height; OD, optical density; pj2*, central wavelength (in cm^{-1}); Δxj*, spectral bandwidth at half-height (in cm^{-1}).*

The above data can be analyzed in terms of spectral and statistical parameters, such as number of Gaussian components, peak height percentages (%OD), half-height bandwidth (Δxj) and peak height (pj). These parameters for all samples are summarized in tab. 5.

Amide bands centered between 1658 and 1652 cm^{-1} are generally considered characteristic of α-helical structures (Bramanti and Benedetti, 1996). In the solid state, the frequency of this component shifts towards higher values (1658-1660 cm^{-1}), due to the stabilization of helical structures by water molecule removal. The components at 1644-1648 cm^{-1} are associated with polypeptide fragments in disordered (random) conformation. Finally, wavelength around 1694, 1692, 1680, 1638 (in solution) and 1614 cm^{-1} are attributed to turns of different types which cannot be easily distinguished (Bramanti and Benedetti, 1996, and references therein). The secondary structures for all the samples are summarized in tab. 6.

Sample	α-helix	β-sheet	random	turn
Solution	30	20	27	23
film 1.7g/l	30	22	18	30
film 0.85g/l	31	20	22	27
film 0.43g/l	22	10	48	47
film 0.21g/l	19	1	45	35

Table 6. *Secondary structure component percentages of cp20 in solution and in solid state, as obtained from the FT-IR spectra of fig. 43 and 44. The random structure includes the extended coil component.*

There is an interesting trend of secondary structure components of solid state samples in relation with the concentration of starting solutions; ordered structures (α-helix and β-sheet) increase with the concentration, as shown in fig. 46. The percentage values found for the solution 5-7 % w/w are consistent with the secondary structure that might be expected for a solid state sample obtained from a concentration of about 0.75 g/l.

It was not possible to measure FT-IR solution spectra at different concentration, because the limit of sensitivity (5 % w/w) was also the maximum concentration before precipitation. CD spectra did not change when the analyzed solutions were diluted 1:2 or 1:3. This was the maximum analyzable range between the limit of sensitivity and the limit of absorbance at 190 nm, which must be below 1 in order to analyze the results with the single value decomposition algorithm. While experiments are in progress to analyze the CD of precipitated cp20 resolubilized in proper detergents, from these preliminary results it appears that the aggregation phenomenon of cp20 may be due to complex mechanisms of phase transition.

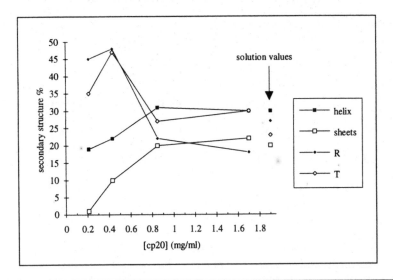

Local protein conformation can be also predicted on the basis of the primary structure by sequence pattern matching (as described in Materials and Methods). Theoretical calculation results of the SOPMA computation algorithm for protein secondary structure are reported in fig. 47.

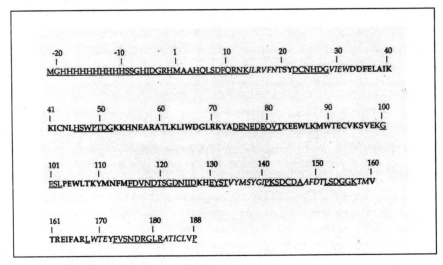

The comparison of cp20 secondary structure results from the three methods is summarized in tab. 7.

	alpha	beta	other
CD	0.33	0.19	0.48
FT-IR	0.30	0.20	0.50
SOPMA	0.36	0.16	0.48

Table 7. *Cp20 secondary structure. CD "beta" structures are the sum of antiparallel and parallel β-sheets; "other" is the sum of turns and random coils. FT-IR "other components" are random structures, turns and extended α-helix.*

These results are in very good agreement with each other; even the β-sheet and the less ordered structures, which are usually much harder than the α-helix component to predict with CD and theoretical computations based on sequence-matching algorithms, correspond closely. The three methods not only confirm and support each other, but also yield slightly different information; while CD discriminates between parallel and antiparallel β-sheet, FT-IR indicates the amount of extended α-helix (tab. 5). Computer-based predictions, while dividing the secondary structure in only three components, also evaluate the location of ordered structures in the aminoacid chain. For example, it is possible to expect that natural cp20 would show a higher content of α-helix and β-sheet than the cloned protein, as the fused oligo-histidine tail is in random (or turn) structure.

The effect of Ca^{2+} binding on cp20 structure was studied by circular dichroism. Free calcium concentration can be accurately controlled by EGTA or other divalent cation chelator, such as EDTA. These compounds cannot be used in FT-IT analyses because of the strong absorption of their carboxylic groups in the Amide regions. The addition of Ca^{2+} to cp20 solution did not strongly affect CD spectra, while the addition of EDTA or EGTA had a dramatic effect on the spectral shape. This finding suggest that Ca^{2+} binding triggers a conformational transition in cp20's secondary structure, but the cloned protein retains calcium along the purification from bacteria. In fact, in contrast with the natural protein purification procedures, no EGTA or EDTA are added along the isolation process. Removal of calcium caused a reduction of the dichroic band at 224 nm with respect to the one at 208 nm, and a parallel reduction of the band at 194 nm (fig. 48). These features are consistent with a decrease of the α-helical content.

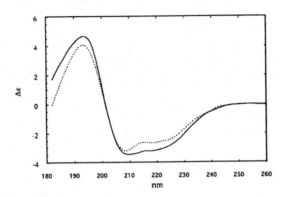

Figure 48. *CD spectra of cp20 at 100 μM (__) or 0.01 μM (...) free calcium concentration. 20 μl of 10 μM cp20 were analyzed in 5 mM trisCl, pH 7.5*

pCa	H	A	P	T	R	H/B
3.40	0.36	0.13	0.01	0.20	0.30	2.57
4.00	0.36	0.13	0.01	0.20	0.30	2.57
4.52	0.35	0.13	0.01	0.20	0.31	2.50
5.00	0.34	0.13	0.01	0.20	0.32	2.42
5.67	0.34	0.14	0.02	0.20	0.30	2.12
6.02	0.32	0.15	0.02	0.20	0.31	1.88
6.29	0.31	0.15	0.03	0.20	0.31	1.72
6.50	0.30	0.15	0.03	0.21	0.31	1.67
6.70	0.31	0.16	0.04	0.19	0.30	1.55
7.02	0.28	0.18	0.03	0.20	0.32	1.33
7.51	0.29	0.17	0.05	0.19	0.30	1.32
8.00	0.26	0.18	0.05	0.19	0.31	1.13
8.45	0.27	0.17	0.05	0.20	0.31	1.23
8.87	0.27	0.18	0.04	0.20	0.31	1.23
9.37	0.26	0.17	0.05	0.19	0.32	1.18
9.77	0.28	0.18	0.05	0.19	0.30	1.22

H=alpha helix   T=turn
A=antiparallel beta sheet   R=random coil
P=parallel beta sheet   B=A+P

Table 8. *Ca²⁺ titration of cp20 secondary structure decomposed from CD spectra. Free [Ca²⁺] is reported as the antilogarithmic value pCa.*

A complete titration of the calcium-induced conformational transition of cp20 was then performed, in the range of 398.5 to 0.00017 μM free calcium. The complete decomposed results are reported in tab. 8.

The alpha helical content increased from 0.28 to 0.35 (25%) upon calcium binding, with a corresponding decrease of the beta structures: 0.18 to 0.13 for the antiparallel sheet (28%) and 0.05 to 0.01 for the parallel sheet (80%). In contrast, turns and coils had constant percentages of 20 and 31%, respectively.

The extent of these changes is consistent with studies carried out on other EF-hand calcium binding proteins (e.g., Amburgey et al., 1995). Changes in the amount of turns or random coil structures with calcium binding are not usually reported, while the α-helix increases (e.g., Durussel et al., 1993; for a recent and extremely complete review on EF-hand calcium-binding proteins, see Sheterline et al., 1995) or decreases (e.g., Zhao et al., 1994) in different proteins without an apparent correlation with other characteristics.

In some experiments, the calcium binding transitional effect overlapped with the instability effect described in tab. 4, in particular when samples where thawed at different times before the CD analysis. In general, a constant value of random coil was considered a key parameter to assess the integrity of cp20 (as the content of random coil remarkably increases when the structure becomes unstable). Nonetheless, both α-helix and β-structures slightly fluctuated through the calcium titration, with a clear issue of instability (as the ordered structures increased or decreased together).

Any attempt to fit α and β structures independently failed to give a satisfactory correlation to $[Ca^{2+}]$. Therefore, the ratio of α-helix over parallel and antiparallel β-sheet amounts was considered a structural marker of the calcium-induced conformational transition, and the data were fit to a sigmoidal, first-order equilibrium equation (fig. 49).

It is worth reminding that cp20 binds two moles of calcium per mole of protein, as assessed by biochemical experiments (chapter 2).

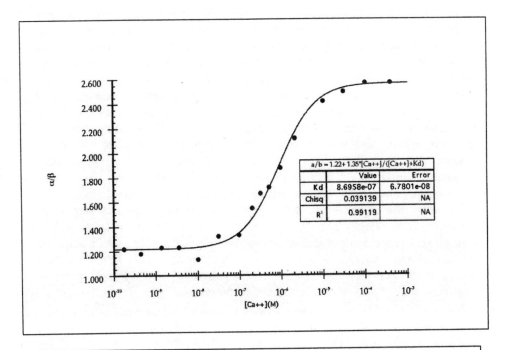

Figure 49. *Effect of Ca²⁺ binding on the relative composition of α-helix and β-sheet (as a sum of parallel and antiparallel structures). The fitting to a first-order dissociation equilibrium (cp20Ca²⁺ ↔ cp20 + Ca²⁺) had a square correlation factor of 99.1% (χ² = 0.039) and an estimated equilibrium constant of 870±70 nM. Plateau values of α/β for low and high calcium conditions were 1.22 and 2.57, respectively.*

An attempt to fit the above curve with a second order equilibrium (with the additional transition at $[Ca^{2+}]$ between 10^{-6} and 10^{-7} M) resulted in a correlation factor of only 56%. It was also impossible to fit the above data with equations assuming a cooperative effect between the two binding sites. These results indicate that only one of two binding sites can modulate cp20 structure. Since the constant evaluated for biochemical experiments showed a higher affinity than that obtained by CD titration (40 nM vs 1 μM), it is reasonable to assume that the first molecule of Ca^{2+} bound to cp20 does not trigger any conformational transition, while the second one induces the secondary structure switch.

Calcium-binding proteins are usually divided in two categories, namely buffering proteins and modulatory proteins (Chazin, 1995). Buffering proteins have high affinity for calcium and contribute to maintaining a low intracellular concentration of calcium by sequestrating the ion. These proteins do not usually undergo a conformational transition upon calcium binding (Sheterline et al., 1995), and their affinity for Ca^{2+} is high at any calcium concentration (i.e. at either "resting potential" or "action potential" in neurons). Modulatory proteins, in contrast, have a lower affinity for Ca^{2+}, depending on the cellular system in which they act (Miller, 1995), and only bind the ion in particular conditions of high Ca^{2+} concentration (e.g. at action potential but not at resting potential in neurons). They are called modulatory because calcium binding triggers a conformational change that allows them to interact with (and modulate) other cellular entities (proteins, receptors, nucleic acids etc.). In other cases, their action is inhibited by calcium binding. Whatever the action mechanism is, calcium binding affects their modulatory activity via a conformational transition. In other words, modulatory proteins are intracellular Ca^{2+} signaling devices. Many EF-hand proteins are modulatory (Sheterline et al., 1995), but some of them are buffering (e.g., Johansson et al., 1993). There are several examples of EF-hand proteins with at least two calcium binding sites, and among the sites of the same protein, some are modulatory and some are buffering domains (Dell'Angelica et al., 1994; Lundberg et al., 1995. For particular examples of squid optic lobe EF-hand proteins, Sheterline et al., 1995). This might indeed be the case for cp20.

It is interesting to observe that, while the cp20 buffering site has an affinity constant (40 nM) that is likely to bind calcium also at neuron resting potential, the transition range of the modulatory domain (10^{-5} to 10^{-7} M) is considered to be the physiological calcium concentration gap between resting and action potential in neurons (Shao et al., 1996, and references therein). In practice, it is possible to speculate that *in vivo* cp20 would assume a high-β conformation at the neuron's resting conditions (when Ca^{2+} concentration is around 10-100 nM), and would switch to a high-α conformation upon firing (i.e. at the neuron's excited state, with an intracellular calcium concentration of 10-100 μM).

The effect of several parameters was studied on the calcium-induced conformational transition of cp20. In vitro CD studies were performed by measuring the α/β ratio (i.e. by measuring the CD spectrum, deconvoluting the secondary structure and calculating the ratio between α-helix and the sum of parallel and antiparallel β-sheets) in different conditions of Mg^{2+} concentration[1], pH and ionic strength. For each analyzed parameter, three [Ca^{2+}] points (10^{-7}, 10^{-6} and 10^{-5} M) were recorded, i.e. basal low-calcium, midpoint of the transition and plateau high-calcium, respectively. The data are summarized in fig. 50.

---

[1] Both Ca^{2+} and Mg^{2+} can be buffered at the same time with EGTA. The free Ca^{2+} concentration is also modified by the pH, so when the pH is varied, calcium must be rebuffered. For the equilibrium equations, see the software "EGTA" at the anonymous ftp site las1.ninds.nih.gov/pub/dos.

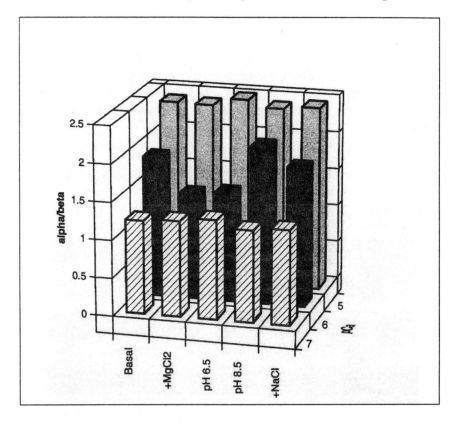

Figure 50. *Effect of different parameters on the calcium-induced conformational transition of cp20. When MgCl₂ was added, the final free Mg²⁺ concentration was 0.5 mM. The pH was buffered by 15 mM trisCl. When NaCl was added, the total ionic strength was 150 mM. The x axis represents antilogarithmic free calcium concentration, the y axis shows the parameters, the z axis indicates α/β.*

The α/β values at pCa 5 and 7 were not affected by any of the tested parameters, indicating that Mg²⁺, pH (within the range 6.5-8.5) and ionic strength did not themselves trigger a conformational transition of cp20. In contrast, the α/β ratio of the EF-hand calcium binding protein Calmodulin is changed by ionic strength even more dramatically than by Ca²⁺ itself, as assessed by similar CD experiments (Hennessey et al., 1987). From the α/β values at pCa=6, in addition, it is possible to determine which parameters affect the Ca²⁺-induced transition equilibrium of cp20.

In particular, Mg^{2+} slightly inhibits the binding effect[2], as does a more acidic pH. In contrast, basic pHs increase the transition effect, as expected and as reported for other proteins (Ames et al., 1995a; Akke et al., 1995). In fact, several acidic residues in the active site can be protonated/deprotonated; the net charge of the binding domain and the coordination efficacy of carboxylic oxygen atoms are affected. The high ionic strength-sample was not significantly different from basal conditions.

The influence of calcium binding on secondary structure was also studied by intrinsic fluorescence spectroscopy, at an excitation wavelength which should mainly act on tryptophan residues. Only one experiment was performed, because the intensity of the signal was too low to allow the analysis of diluted samples. The samples recovered from the previous CD experiments were recovered, diluted (concentrated small volumes are needed for CD, while diluted large volumes can be analyzed by fluorescence) and rebuffered. Cp20 presented a maximum emission at 338 nm (typical of tryptophans), and the signal remarkably increased upon calcium removal (fig. 51).

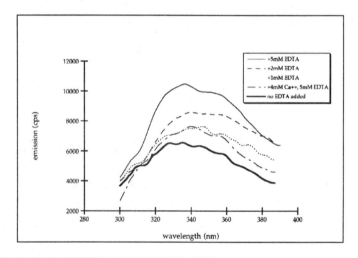

Figure 51. *Effect of Ca^{2+} removal on cp20 intrinsic fluorescence (average of three measurements). [cp20]=400 nM in 1 mM Ca^{2+}, 20 mM trisCl (pH 7.5), V=3 ml, cell pathlength 1 cm, excitation at 279 nm. The thick line is basal (calcium-bound) cp20. EDTA was added to 1 mM final (dotted line), 2 mM final (broken line), 5 mM final (full line). [Ca^{2+}] was then restored to 5 mM total, and the signal decreased to the centered line (close to the dotted line).*

---

[2] It is not possible to discriminate between an actual reduction of Ca^{2+} binding and a decrease of its transitional effect (or a combination of the two). If the physiological modulation of cp20 is mediated by the conformational transition, however, the two mechanisms are equivalent (see also the discussion in Miura et al., 1994).

In order to verify that the fluorescence signal increase was not due to photodegradation (the solution is excited at 279 nm for over an hour at room temperature), an internal control was performed, consisting of increasing the free calcium concentration back to an intermediate value. The fluorescence intensity returned to the correspondent level.

There is not a general rule to predict the variation in intrinsic fluorescence of EF-hand proteins upon calcium binding, as it either increases, decreases or remains constant (e.g., Pottgiesser et al., 1994; Dell'Angelica et al., 1994). In general, only modulatory protein show a change of their fluorescence, corresponding with a structural transition (Sheterline, 1995). A decrease of the signal with calcium binding is often consistent with an increase in the α-helix content (Durussel et al., 1993). Therefore this experiment confirms with an independent technique that Ca^{2+} triggers a transition in cp20 secondary structure. It is possible to speculate that the folding of cp20 is tighter in the calcium bound-form (as tryptophan residues are less exposed), which is also consistent with an increased overall globularity (due to the higher α-helical content).

An increase in the fluorescence signal might also indicate a change in the overall polarity of the protein, as the aromatic, hydrophobic aminoacidic residues have a different disposition to the solvent. This could be of particular interest for cp20, which is known to translocate to the membrane upon neural stimulation (Nelson and Alkon, 1995). Thereafter, a chromatographic experiment was set up to verify this possibility; phenyl-sepharose is often used as a reverse-phase resin with high sensitivity to polarity changes. In general, high ionic strength promotes protein affinity for this resin, encouraging non-ionic and aromatic (π-π) interactions. A variation of the retention time on phenyl-sepharose upon calcium binding was also suggested as a test to discriminate between modulatory and buffering EF-hand proteins (Donaldson et al., 1995, and references therein). Cp20 proved to have a very high affinity for phenyl-sepharose[3], and even at a low ionic strength it was impossible to elute it from the stationary phase without the use of urea as a denaturant. The presence or absence of calcium did not change the retention time (fig. 52).

---

[3] The low polarity of cp20 is consistent with the phenomenon of precipitation from aqueous solutions frequently observed during the purifications from both squid optic lobe and E. coli, and also with the results obtained by solid state FT-IR (fig. 44). Actually, in squid optic lobe tissue, about 50 % of the protein is membrane-bound, and almost the totality of rabbit brain cp20 is in the membrane fraction. Cp20 is a cytosolic protein which can translocate to the cellular membrane (Nelson and Alkon, 1995), where it interacts with the channels; a high affinity for phenyl-sepharose is thus expected.

In conclusion, calcium binding does not seem to mediate directly cp20's compartmentalization (cytosol or cellular membrane), although more experiments are necessary to confirm this preliminary indication.

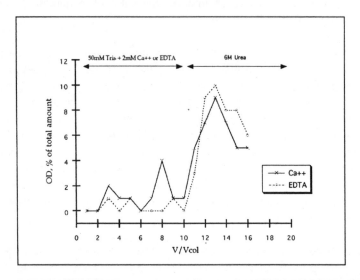

Figure 52. *Elution of cp20 on a low-pressure phenyl-sepharose chromatographic column with and without Ca^{2+}. Detection with pAb dot-blot. V$_{col}$ = 2ml, loaded sample: 112 ng/85 μl. Flow rate: 0.5 ml/min.*

In order to rule out the possibility that the above described calcium-binding and second structural features were peculiar of the cloned protein, due to the oligo-histidine fused tail, and not of the natural squid protein, a catalytic reaction was implemented to remove proteolitically the extra aminoacidic chain. The sequence inserted between the oligo-histidine tail and the beginning of natural cp20 is a consensus for Factor X, an endoprotease involved in the blood coagulation pathways.

Upon oxidative activation, Factor Xa (activated factor ten) transforms the protein prothrombin in active thrombin by removing an aminoacidic tail next to the consensus sequence, thus starting the coagulation process (Aurell et al., 1977, and references therein). The reason for the great utility of this protease is the extreme specificity for the consensus sequence (Ile-Glu-Gly-Arg or Ile-Asp-Gly-Arg). In addition, this sequence is extremely rare in natural proteins (Nagai and Toegersen, 1984). Therefore, by introducing the consensus aminoacids in a cloned protein with common molecular biology procedures, it is possible to remove selectively the entire portion of the protein between the above sequence and the N-terminus.

Cloned cp20 was reacted with commercial bovine activated Factor Xa in standard conditions (see Materials and Methods), and after two hours at room temperature the reaction was stopped by dilution with protease inhibitor cocktail (which blocked Factor Xa activity) and two consecutive passages through a 50 kD cut-off ultrafiltration membranes (C50). While cp20 was almost entirely recovered, Factor Xa (MW = 47000) was completely retained in the ultrafiltration device. The histidine tail removal reaction had an overall yield of 50%, as assessed by silver stained SDS-PAGE and western blot (fig. 53). If the reaction mixture was incubated for longer time, some non-specific proteolytic degradation bands began to become visible.

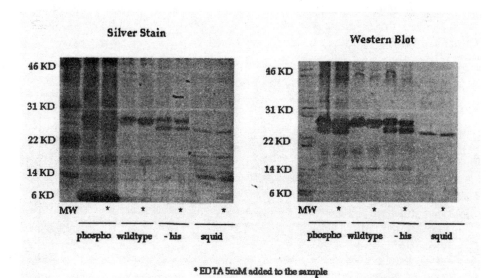

Figure 53. *Silver stain (left) and western blot (right) SDS-PAGE analysis of the crude reaction products of purified cloned cp20 with Factor Xa and PKC (see text below). First lane: MW markers. Second and third lanes: PKC reaction product. Fourth and fifth lanes: purified cloned protein without modifications. Sixth and seventh lanes: Reaction product with Factor Xa. Last two lanes: control of purified natural squid optic lobe cp20 (prepared as described in chapter 1). Samples were loaded in duplicate, and EDTA was added in the lanes marked with an asterisk.*

As expected, the lower band in the reaction product mixture is at the same height as the natural squid optic lobe cp20, because the additional aminoacidic tail has been removed. The higher band is unreacted protein. The molecular weights of the two proteins (25 kD and 22 kD) are consistent with the calculated values and with previously described results (chapters 1 and 2). The presence of a minor impurity at 18 kD, corresponding to the main non-specific proteolytic fragment of cp20, is also observed in both cloned and natural proteins.

The reaction mixture was then repurified by IMAC-HPLC. Unreacted cloned cp20 was retained in the column due to the high affinity of the oligo-histidine tail for Ni^{2+}, while proteolyzed cp20 eluted soon after the void volume (fig. 54).

An SDS-PAGE analysis of the repurified product showed a single band on silver staining, at the molecular weight characteristic for cp20 without the oligo-histidine tail (22 kD; fig. 55).

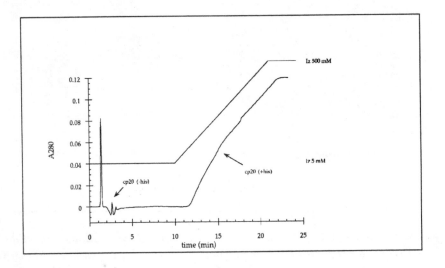

Figure 54. *Ni^{2+}-IMAC HPLC elution of the reaction mixture of cp20 with Factor Xa after C50 passage. The arrows indicate the small peak of proteolyzed cp20 (without the histidine tail) after the large void volume peak and unreacted cloned cp20 (with the histidine tail) eluting as a broad peak on the imidazole (Iz) gradient.*

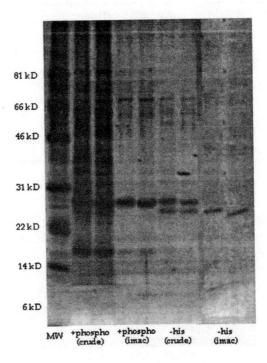

Figure 55. *Silver stained SDS-PAGE of Factor Xa- and PKC-reacted cp20 (for the reaction with PKC, see below in the text) before and after the IMAC purification. Samples were run in duplicate, and EDTA was added to the right lane of each pair.*

The second covalent modification that was induced on cloned cp20 was the PKC phosphorylation of the residue thr_{61}. The reaction carried out by commercial PKC was however incomplete, as assessed by externally sampling with ^{32}P (tab. 9); it is likely that PKC itself had a major impurity of phosphatase activity, as the results obtained with cp20 were very similar to those obtained with the positive control, histone. The kinetics of the phosphorylation showed in tab. 9 also indicates the presence of phosphatase activity.

Time (min)	cp20	Histone	BSA
0	3741	2715	7661
5	6809	19522	3409
15	14304	33210	7096
30	54723	36411	1254
60	27414	67361	2156
120	9506	14377	5932
no PKC, 30	8409	19112	11421

Table 9. In vitro α-PKC-induced phosphorylation of cp20 measured by an external pilot reaction with ^{32}P-ATP. Under standard PKC-catalyzed conditions, histone covalently binds 1 mole of phosphate per mole of protein (positive control), BSA none (negative control). Data are expressed in disintegration per minute (DPM).

From the stoichiometric conditions of the reaction, and the activity of ^{32}P (see Materials and Methods), it is possible to evaluate that approximately 10-30 % of cp20 was phosphorylated after 30 minutes.

The reaction mixture presented several impurities on silver stained SDS-PAGE (fig. 53), in particular high MW bands due to commercial PKC impurities and its degradation products, and low MW bands due to multilamellar vesicles. The mixture was therefore repurified on Ni^{2+}-IMAC-HPLC, and the resulting product showed a single band on gel (fig. 55; PKC and all the other components were eluted in the void volume).

Oligo-histidine-removed cp20 and phosphorylated cp20 were tested for their calcium-binding activity in comparison to the unmodified protein (fig. 56). Only a few points of the titration were measured, because of the low amount of samples.

The removal of the oligo-histidine tail from the aminoacidic sequence does not affect the calcium-induced conformational equilibrium, suggesting that natural squid optic lobe cp20 might also undergo the same secondary structure switch.

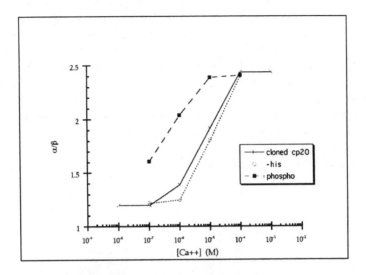

Figure 56. *Titration of the variation of cp20 secondary structure (expressed as a ratio between the relative contents of α-helix and β-structures) upon calcium binding (same conditions as in fig. 49). Broken line, "phospho" (partially phosphorylated, repurified cp20), fitting the experimental points (filled squares); dotted line, "-his" (repurified cp20 without histidine tail), fitting the experimental points (empty circles). Full line, cloned cp20 (unmodified) fitting the points (crosses) from fig. 49.*

In contrast, a clear difference is observed with the phosphorylated cp20. Unfortunately, the small number of points does not allow one to discriminate between a reduction of the equilibrium constant by one order of magnitude and a different plateau in the apo-form. It should be also considered that the phosphorylation attempt was not completely successful, and only a small fraction of the protein is phosphorylated (thus the effect of a complete phosphorylation might be more dramatic). From these preliminary observations it is tempting to speculate that phosphorylation blocks, or stabilize cp20 in a conformation close to the calcium bound form. A similar mechanism of interaction between calcium binding and a covalent post-translational modification, with an extremely meaningful effect on the protein's signaling activity, was recently described for the myristoylation of the retinal protein Recoverin (e.g. Ames et al., 1995, and references therein).

From a careful analysis of fig. 53 and fig. 55, it is possible to observe that in each sample, the addition of EDTA (and hence the removal of Ca^{2+}) slightly increases the electrophoretic mobility (i.e. the apparent MW is lower). Although the harsh and denaturing conditions of SDS-PAGE are not appropriate to investigate conformational properties of cp20, in the case of modulatory EF-hand proteins this variation of the electrophoretic mobility is usually interpreted as a further clue that Ca^{2+} binding indeed affects the secondary structure (Lu et al., 1994). A non-denaturing gel electrophoresis experiment was therefore set up to study this phenomenon in more detail in native conditions. All samples (pure cloned cp20, partially phosphorylated cp20, cloned histidine tail-removed cp20 and natural squid optic lobe purified cp20) were loaded with either $CaCl_2$ or EDTA and run in a non-SDS polyacrylamide gel. The resulting blot was electrtransferred on a nitrocellulose membrane and developed as a western blot (fig. 57).

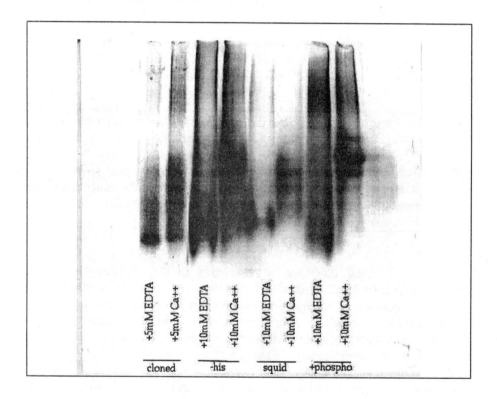

Figure 57. *Native gel electrophoresis western blot of cp20 proteins (unmodified cloned cp20, lanes 1 and 2 from the left; Factor Xa-modified cp20, lanes 3 and 4; squid purified cp20, lanes 5 and 6; PKC-modified cp20, lanes 7 and 8). The blot was developed with anti-cp20 polyclonal antibody.*

A much more dramatic effect is observed in native conditions in comparison to that shown in the denaturing gels. The non-denaturing PAGE does not yield well-focused bands, because a conformational equilibrium is present and the protein is a mixture of different structures. However, the sequestration of calcium clearly increases the electrophoretic mobility of all samples, as quantified in fig. 58.

Partially phosphorylated cp20 did not behave differently than the other samples. This would suggest that phosphorylation acts on the equilibrium of apo-protein - calcium-bound protein, rather than on the absolute conformation (only extreme conditions of calcium concentrations can be studied by PAGE); it should be underlined however that the PKC reaction was not completely satisfactory, and these results should therefore be considered preliminary.

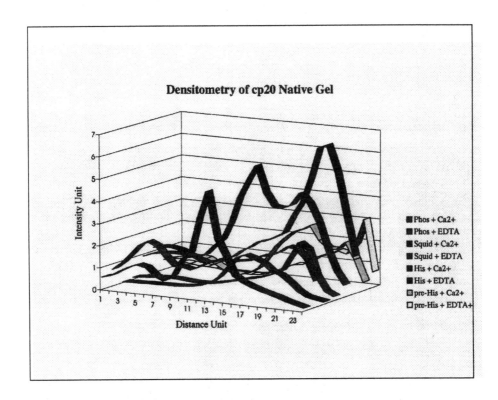

Figure 58. *Quantitative scan densitometry of the western blot in fig. 57. For each pair of samples, the Ca^{2+} loaded protein had a slower run (shorter distance from the beginning of the gel). The electrophoretic mobility is similar for all samples in the same Ca^{2+}/EDTA conditions.*

These experiments provide a totally independent, non-spectroscopical evidence of the Ca^{2+} induced conformational transition of cp20. In addition, since PAGE apparent molecular weight (i.e. electrophoretic mobility) depends on the shape of the protein, these results indicate that cp20 is more extended and elongated in the apo form, and acquires a more globular, compact and bulky shape upon calcium binding. This finding is in full agreement with the CD results (higher α-helix with calcium) and the fluorescence experiments (higher tryptophan exposure without calcium). In addition, PAGE experiments require a minimal amount of sample (50 ng per lane), and a direct analysis of the squid optic lobe protein was therefore performed (the natural protein was not pure and abundant enough to be characterized spectroscopically). The equivalence between the natural protein and the oligo-histidine tail-removed cloned cp20 was thus established, and the same behavior of cloned and natural cp20 upon calcium binding suggested.

In conclusion, a complete secondary structure characterization of cp20 has been achieved. Three independent methods (CD, FT-IR and SOPMA) yielded a consistent composition of secondary structures. Cp20 conformation was not stable for long time *in vitro*, with a general increase of random coils over α-helix, β-sheet and turns. In addition, an aggregation phenomenon was observed, which caused a slight increase of α-helix and a decrease of the β-sheet, and more generally an increase of random coils (including turns) with respect to overall ordered structures (when comparing solid state with highly concentrated solutions). This variation of secondary structure upon phase transition, however, was remarkably dependent on the concentration of the starting solution, suggesting that aggregation might be mediated by a complex mechanism. Furthermore, a remarkable conformational switch was observed and characterized upon calcium binding. An increase of α-helix and a decrease of both parallel and antiparallel β-sheets accompanied the complexation of Ca^{2+} at the low-affinity, modulatory EF-hand site. This switch changed cp20 shape (elongation and exposure of aromatic residues), but not the overall protein polarity. From the experiments presented in this chapter, no difference could be observed in the behavior of the natural squid optic lobe cp20 and the cloned, bacterial protein.

CHAPTER 4

Imaging PKC activation in living cells

The characterization of cp20's secondary structure initiated with this thesis opens more mechanistic questions on how exactly cellular signals, such as an increase in the calcium concentration or the activation of PKC, relate to cp20 conformational changes in affecting the potassium channel activity (the "memory trace"). In particular, the fact that the phosphorylation by PKC induces the translocation of cp20 into the cellular membrane (Nelson and Alkon, 1995), where potassium channels are, leaves an open question; since PKC itself translocates into the membrane upon activation (by calcium), where is cp20 phosphorylated? In the cytosol, by a partially activated PKC, in which case the phosphorylation itself could drive cp20 into the membrane?, or in the membrane, where active PKC should be? In the second case, it has to be assumed that cp20 is normally in an equilibrium of compartmentalizations (between the membrane and the cytosol), and phosphorylation of the membrane-cp20 fraction drastically moves the equilibrium.

The most direct way of answering these questions is to set up an *in vivo* imaging system to view the intracellular position of both PKC and cp20. Confocal microscopy is the technique of choice for this experiment, but it requires a fluorescent probe that specifically binds the proteins (in order to mark them in a complex matrix such as the cellular milieu). Cp20 does not have a specific ligand, since GTP and calcium also bind to many other proteins. One possibility would be to purify cp20, to label it covalently with some fluorescent probe, and to microinject it into the cell. Another possibility is to conjugate cp20's polyclonal antibody with a fluorescent antenna. In this case there would be the problem of how to control the loss of cp20 functionality. While experiments are in progress to assess these possibilities and to test alternative strategies, PKC has a natural, specific ligand, the family of phorbols. Commercial phorbols labeled for fluorescence spectroscopy and microscopy are also available. They have a very high diffusion rate into the membrane, and can be absorbed intracellularly without microinjection. However, imaging PKC in living neurons presents a number of problems. For instance, conditions must be found to preserve the neuron's health all through the experiment, for several hours on a microscope slide. In addition, the exciting light beam induces photodegradation and bleach of intracellular molecules. This effect causes a strong autofluorescence (i.e. the cell emits a signal independent from the probe), which interferes with the experiment. In order to set up an imaging protocol and to assess these preliminary problems, a non-neural cellular model was chosen, the sea urchin egg. Sea urchin eggs are suitable for confocal microscopy because they have very little autofluorescence, they are big (up to 100 μm diameter) and round shaped, thus easy to image, and they are quite resistant. Additionally, like neurons, eggs are excitable cells, that can be activated by a physiological stimulus (the addition of sperm and the fertilization).

It was observed that developing systems (such as a fertilized egg) show an activated biochemical pathway that shares some characteristics with stimulated neurons, and in particular with synaptic plasticity. For instance, PKC translocation was proved to occur during fertilization (Bloom et al., 1995), and is surely involved in the cellular mechanisms of associative conditioning (for a review, see Alkon, 1989). In general, most metabolic pathways active in CNSs and neurons to control synaptic plasticity are also present in other cellular systems of different species, being conserved through evolution for their importance. Cellular mechanisms of learning and memory evolutionarily developed from non-neural plastic systems specialized in the conservation of genetic and biological memory, such as developing and dividing eggs (Dudai, 1989; Stabel and Parker, 1991). In this perspective, synaptic plasticity is only a particular case of a widely diffused cellular and biochemical plasticity (Ashendel, 1985). Even the cp20 pathway is not uniquely present in neuronal systems, but has been characterized in different developing cells such as fibroblasts (Kim et al., 1995) and sea urchin eggs (Nelson et al., 1996a). The PKC pathway involves several activation steps which start from a first messenger, i.e. extracellular signals (hormones, electrical potential, neurotransmitter) and end with the phosphorylation of a substrate (such as cp20, potassium channels, B50 or neurogranin, see also part C of the appendix). In particular, PKC acts as the second messenger, i.e. a crucial intracellular nexus, for the phospholipase (PL) cascade (Olds and Alkon, 1993); when a particular extracellular receptorial proteins is activated by the first messenger, a transmembrane conformational change occurs, and information is transduced into the cytoplasmatic side. In case of PL, an enzymatic activity is stimulated, which starts the phospholypidic catabolism. From triglycerides, two species are then formed: diacylglycerol (DAG), a direct activator of PKC, and phosphoinositol-triphosphate (PIP_3), which indirectly activates PKC by acting on endoplasmatic reticulum (the main intracellular calcium storage) to release Ca^{2+}, another activator of PKC. Moreover, PIP_3 undergoes further degradation and other phospholypidic effectors of PKC are produced. Such a chain of events has been demonstrated to be of crucial importance in the precise regulation of phosphate incorporation in PKC substrates (for reviews, see Olds and Alkon, 1993; Nishizuka, 1992). This scheme is shared between neural and non-neural cells.

Over the last thirty years, the sea urchin egg has become an important model system for understanding the role of signal transduction events on the first cell cycle (Bonnell et al., 1994; Cameron and Poccia, 1994; Li et al., 1994; Johnston and Sloboda, 1992; Steinhardt, 1990a and 1990b; Baitinger et al., 1990; Ciapa et al., 1989; Gillot et al., 1989; Shen and Ricke, 1989; Hinegardner, 1975; Ohlendiek and Lennarz, 1995; Hagstrom and Lonning, 1973). Subsequent to fertilization, activation of the sperm receptor triggers the beginning of the phospholipase pathway with a corresponding change in the distribution of intracellular Ca^{2+} stores. Subsequently, the Na^+/K^+ antiporter is activated which leads to elevation of the fertilization membrane mediated by cortical granule fusion with the plasma membrane. The receptor for the sea urchin spermatozoa has been cloned and biochemically characterized (Ohlendiek and Lennarz, 1995); however, G-protein or tyrosine kinase mediated activation of phospholipase C (and possibly A_2) has only been hypothesized as the cause of an IP_3/ Ca^{2+}-dependent Ca^{2+} mobilization (Moore et al., 1994; Moore and Kinsey, 1995). The initial Ca^{2+} transient has both temporal and spatial wave-like characteristics (Steinhardt, 1990a and 1990b; Gillot et al., 1989) and may contribute with diacylglycerols and/or arachidonic acid to the activation of a recently cloned 72.4 kD Ca^{2+}-dependent protein kinase (PKC; Shen and Ricke, 1989; Rakow and Shen, 1994). Other works demonstrated that phorbol esters can mimic the effects of fertilization on alkalinization of the egg (Ciapa et al., 1989). Furthermore, microinjection of PKC pseudosubstrate can at least partially block the alkalinization (Shen and Buck, 1990). Analogous studies on *Chaetopterus* oocytes have also demonstrated a role for PKC in the events subsequent to fertilization, leading to germinal vesicle breakdown (Eckberg and Palazzo, 1992; Bloom et al., 1995; Eckberg and Carroll, 1995; Eckberg et al., 1987; Eckberg and Szuts, 1993; Connor et al., 1992). Catalytic activation of the calcium-dependent isozyme of PKC that is present in relatively high amounts in sea urchin egg would require, in addition to Ca^{2+}, an appropriate biophysical microenvironment that contains phosphatidylserine and the appropriate lipid activator (Zidovetzki and Lester, 1992; Lester, 1992). This condition correlates with PKC association with cellular membranes (Zidovetzki and Lester, 1992; Epand and Lester, 1990; Olds et al., 1989).

While the role of protein kinase C in the first cell cycle has been demonstrated (Zhou et al., 1993; Watanabe et al., 1992; Berridge and Irvine, 1984), to date the evidence for its fertilization-dependent activation has been from *in vitro* biochemical studies which provide little information on the spatial characteristics of the activation events. Here we present *in vivo* imaging data that sea urchin PKC undergoes a rapid and sustained translocation to the vicinity of the plasma membrane immediately after fertilization. Furthermore, we demonstrate the ability of laser-scanning confocal microscopy to quantitatively visualize PKC activation under *in vivo* conditions.

In particular, the fluorescent dye NBD-phorbol acetate was used to visualize the activation of PKC, in living *Lytichinus pictus* eggs during fertilization. This dye interacts directly with PKC as determined using a competitive binding assay. Quantitative image analysis of sequential images from laser scanning confocal microscopy showed a significant reorganization of the signal in the vicinity of the cortical granules and the plasma membrane that began immediately following fertilization and persisted up to 1 hr ($p < 10^{-4}$). At the concentrations employed, the NBD-phorbol dye was not capable of inducing a significant translocation of the fluorescent signal to the membrane, nor did it appear to interfere with the cell cycle. It therefore seems likely that the present *in vivo* results reflect the previously reported *in vitro* activation of protein kinase C immediately subsequent to fertilization. Such an interpretation is parsimonious with the results of parallel subcellular fractionation experiments using an N-terminal polyclonal antibody to sea urchin PKC which showed a significant ($p < 0.037$) translocation of the enzyme from the cytosolic fraction to the membrane fraction 40 minutes subsequent to fertilization. This study supports and extends previous in vitro data suggesting that PKC activation subsequent to fertilization occurs at or near the egg plasma membrane, perhaps in association with arachidonic acid rich cortical granules.

Experiments are in progress to visualize both PKC and cp20 translocation in living neurons and astrocytes.

Materials and Methods

Sea urchins (*Lytechinus pictus* or *Strogylocentrotus purpuratus*) were obtained from Marinus, Inc. (Long Beach, CA, USA) and kept at approximately 15°C in a flow through sea water aquarium at the Marine Biological Laboratory (Woods Hole, MA, USA) or 10°C in an artificial sea water (ASW) aquarium at the NIH (Bethesda, MD, USA), respectively. Gametes were isolated by intracoelomic injection of 0.5 M KCl. For imaging experiments, the jelly layer was removed from the eggs by two brief exposures to pH 6.0 sea water followed by washing in ASW.

In subcellular distribution of ^{3}H-PDBU (tritiated phorbol dibutyrate) binding experiments, multilamellar vesicles (MLVs; 20% phosphatidylserine : 80% phosphatidylcholine) were prepared as previously described (Lester, 1990). Eggs incubated for 40 minutes with either sperm, boiled sperm, or unfertilized eggs, were homogenized in the binding buffer (20 mM trisCl, pH 7.5; 0.1 mM free $CaCl_2$; 2 mM $MgCl_2$; 0.5 mM DTT). The sample was then centrifuged at 100,000 x g at 4°C for 1 hr and the supernatant (cytosolic fraction) was kept at 4°C. The pellet was resuspended in binding buffer containing 0.5% Triton X-100 and incubated at 4°C for 45 minutes. This pellet was then centrifuged as above and the supernatant designated membrane fraction. These fractions were further purified on a commercial fast flow DEAE-Sepharose column (1 ml, Pharmacia, Uppsala, Sweden). The columns were rinsed with buffer (20 mM trisCl, pH 7.5; 0.5 mM EGTA) and then active PKC eluted with the same buffer containing 0.11M KCl. Glycerol (10%, v/v) was added to stabilize activity. Protein amount in each fraction was determined with the bincichonic test (Smith et al., 1985). ^{3}H-PDBU binding was measured in reaction buffer (20 mM trisCl, pH 7.5; 0.5 mM DTT; 0.5 mM EDTA; 3.6 mM $CaCl_2$; 13 mM $MgCl_2$), giving a final free concentration of Ca^{2+} of 0.1 mM and an excess of Mg^{2+}. Reaction buffer was added to 100 μM MLV's and mixed with ^{3}H-PDBU (Amersham, Cambridge, UK, 20 mCi/mmol;

final concentration 20 nM). This reaction mixture was incubated for 5 minutes at room temperature to allow the phorbol ester to partition into the lipids. Bovine serum albumin (BSA) was added to a final concentration of 0.5 mg/ml. An aliquot (100 μl) of sample (either cytosol or membrane) was then added and the reaction mixture (300 μl) was incubated for 30 minutes at room temperature followed by 60 minutes at 4°C. 1 ml of stopping buffer (0.1 mM Ca^{2+}; 5 mM Mg^{2+}; 20 mM trisCl, pH 7.5; 0.5 mM DTT) was added to terminate the reaction. The samples were harvested on phosphocellulose filters and radioactivity determined as previously described (Lester, 1990). Total PKC bound was defined as the amount of ^{3}H-PDBU bound in the absence of any unlabeled phorbol ester. Background was determined by adding excess (10 μM) phorbol myristoate acetate to the conditions used for total PKC bound. Specific PKC bound was the difference between total and background PKC bound.

For the dot blot procedure, membrane and cytosolic fractions were prepared as above. Aliquots (300 μl) of each of the fractions for the three conditions (fertilized, boiled sperm and unfertilized control) were transferred to nitrocellulose in a dot blot apparatus (see chapters 1 and 2). After overnight blocking in TBS-BSA (1%) at 4°C, blots were incubated with a polyclonal primary antibody directed against the N-terminus of a cloned sea urchin PKC as previously described (Rakow and Shen, 1994). Blots were subsequently washed twice for 4 minutes with TBST and once with TBS at room temperature and then incubated with a secondary antibody (goat anti-rabbit IgG; alkaline-phosphatase conjugate) for 1 hr at room temperature. Blots were washed as above and then stained with the Pierce (Rockford, IL, USA) alkaline phosphatase kit. Dot blots were then scanned and analyzed using the M1 (Imaging Research, St. Chatherines ON, USA) software package.

In competitive phorbol ester binding assays, NBD-phorbol acetate (12-α-(12-(N-methyl-N-(7-nitrobenzyl-2-oxa-1,3-diazol-4-yl)amino))dodeconoyl-phorbol-13-β-acetate; Molecular Probes, Corvallis, OR, USA) was tested for its ability to displace ^{3}H-PDBU. The fluorescent phorbol ester was added at the concentrations indicated and allowed 10 minutes to reach equilibrium with respect to partitioning into the vesicles. The reaction was begun by adding purified rat brain PKC (50 ng) and incubating the mixture for 30 minutes at room temperature, followed by 30 minutes at 4°C. ^{3}H-PDBU binding was then done as described in the subcellular fractionation experiments.

The fluorescent probe NBD-phorbol ester was employed in the labeling procedures on the basis of its previously reported specificity for PKC and lipid solubility (Connor et al., 1992). A stock solution of NBD-phorbol ester 12.8 μM was prepared in dimethyl sulfoxide (DMSO). This stock solution was stored in the dark at -20°C until use. For standard labeling, 9.8 μl of stock solution were added to 1 ml of a suspension of eggs in ASW and allowed to incubate for 30 minutes at 15°C. In some of the validation experiments, the incubation time or the concentration of label were the dependent variable. Eggs were then subjected to gentle centrifugation using a hand centrifuge and then resuspended in sea water. Labeled eggs were washed in this manner three times before fertilization and observation procedures. Activation of eggs was induced by the dilution of dried sperm into ASW as previously described (Cameron and Poccia, 1994; Johnston and Sloboda, 1992; Steinhardt, 1990a; Hinegardner, 1975).

Lytichinus pictus eggs were visualized on a Zeiss Inverted Laser Confocal Microscope system using a 20x (1.30 N.A.Neofluoar) objective immediately after sperm activation (Marine Biological Laboratory, Woods Hole, MA, USA). Eight second excitation scans of the fluorescent probes were accomplished via an external 488 nm ArKr laser with a BS568 line filter to create confocal images which were transferred to an imaging system (see below). Simultaneous scanning using the internal HeNe red 647 nm laser line was used to produce bright field images of the eggs. These bright field images were used to assess the state of the egg (health status, fertilization envelope etc.) for each time point. Generally eggs were imaged 30 sec after the addition of sperm, every minute for the first 5 minutes of the experiment and subsequently every 5 minutes for the duration of the experiment. No significant bleaching of the probe was observed over the course of the observations.

Images from the Zeiss laser confocal system were transferred over the internet to an M1 Imaging System (Imaging Research, St. Catherines ON, USA). In order to standardize image analysis, computer-generated transept bars were then placed over the confocal egg image across a diameter (if the cell was spherical) or the major axis (if the cell was elliptical) with a thickness equal to that of the nucleus (approximately 10 μm). If the nucleus was not visible in the confocal section, the transept bar thickness was adjusted to 1/10 of the cell diameter. The position of the transept bar was set to pass directly through the nucleus if the nucleus was visible in the confocal section. Otherwise, the transept bar was placed arbitrarily perpendicular to the y-axis of the image. This image analysis protocol is illustrated schematically in fig. 59.

The imaging system was then used to determine the average pixel intensity (PI) across the width of the transept bar between the left half-height of the first left peak and the right half-height of the first right peak.

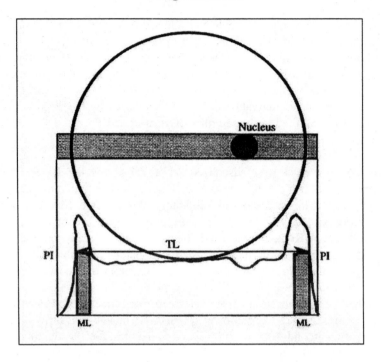

Figure 59. *Schematic illustration of image analysis procedures used to assess fertilization-dependent shifts in the NBD-phorbol ester signal. The fluorescence signal across the transept bar was graphed as pixel intensity (PI; y-axis) versus position (x-axis). The borders of the egg were arbitrarily defined by perpendicular segments to the half maximal heights of the PI graph at the right and left edges.*

The first and last x-axis position corresponding to the average PI were subsequently selected. These points defined the egg's borders and the distance between them was the total length (TL). The membrane locale (ML) was arbitrarily defined as 1/15 of TL, and the external borders of the left and right ML's correspond to the left and right egg borders respectively (fig. 59). The total signal (TS) was defined as the PI within the egg borders and the membrane signal (MS) was defined as the average of the PI for the left and right ML's. The percentage of membrane-locale label (%Z) was then defined by the following equation:

$$\%Z = K(MS-TS)/MS$$

so that %Z=0 when MS=TS and %Z=100 when the entire signal was within the confines of the membrane vicinity(as defined by ML). The annulus constant, K, accounted for the fact that the membrane signal, MS, was a subset of the total signal (TS).

Statistical analyses were carried out using the SYSTAT software package (SAS Inc., Evanston, IL, USA).

Results and Discussion

The NBD-phorbol ester probe showed the ability to bind to PKC in the 5-30 nM range as assessed by its ability to displace ^{3}H-PDBU binding to purified rat PKC (fig. 60). The inhibition curves show IC50s that demonstrate a slightly higher affinity of the fluorescent probe for PKC (Kd ~ 8 nM) than that of PDBU itself (Kd ~ 20 nM).

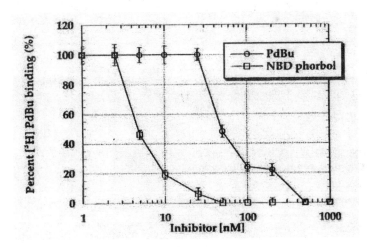

Figure 60. *Fluorescent probe validation experiments: competition of NBD-phorbol ester and cold PDBU against the binding of ^{3}H-PDBU to PKC.*

Image analysis of %Z as a function of concentration of the molecular probe[1] was performed on unfertilized *L. Pictus* eggs and the results are shown in fig. 61. Increasing concentrations of NBD-phorbol ester probe caused a monotonic increase in TS in combination with a sigmoidal increase in %Z (with an apparent constant similar to that of ^{3}H-PDBU dissociation). For subsequent fertilization experiments an NBD-phorbol ester concentration of 25 nM was used. At this concentration only about 5% of the signal was associated with the membrane locale, and any further increase in %Z could only be attributed to physiological changes within the egg.

In order to ensure that the observed signal was due mainly to the PKC-bound probe, two blanks (egg autofluorescence and buffer background) were compared to sample fluorescence (fig. 62).

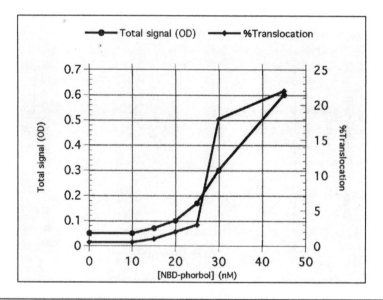

Figure 61. *Graph of phorbol ester label translocation (%Z) and total signal (TS) as a function of NBD-phorbol ester concentration in unfertilized L. Pictus eggs. At 25 nM the compound has a maximum probe efficiency without affecting PKC activation.*

---

[1] The fluorescent antenna NBD is linked to a phorbol moiety to guarantee high affinity and specificity to PKC. However, the activating effect of phorbol (which might cause translocation of PKC) must be taken into account.

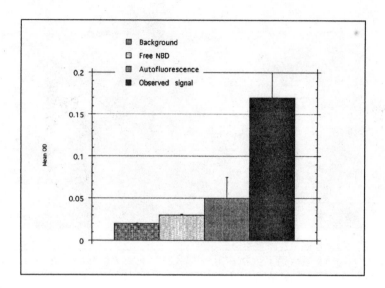

Figure 62. *Comparison between intracellular NBD fluorescence intensity and blank signals. Remarkably, free (extracellular) NBD has a very low signal, because its fluorescence is quenched by polar environments (such as water).*

Cell autofluorescence and background signal did not variate significantly in the course of several hours. Furthermore, the free NBD signal is negligible compared to that of the protein-bound probe. Therefore any signal changes could be attributed to PKC translocation.

A typical fluorescent imaging experiments of *L. Pictus* egg fertilization is shown in the sequence of images of fig. 63 (a-f).

Corresponding bright field images clearly show both the general intact status of the eggs and the elevation of the fertilization envelope (fig. 64).

Quantification of the total signal (TS) for fertilized eggs showed no significant changes over the course of the experiment (fig. 65). This means that the fluorescent probe did not undergo any significant photobleaching process.

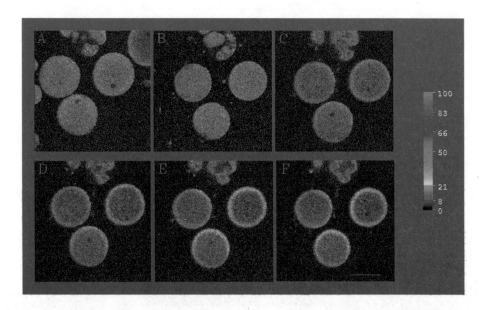

Figure 63. *Time series of confocal images of 3 eggs subsequent to fertilization, showing the long lasting shift in the NBD-phorbol ester signal to the membrane-locale. Scale bar is 110 µm. Time points: A) before sperm addition; B) +30 seconds; C) +10 minutes; D) +20 minutes; E) +40 minutes; F) +130 minutes.*

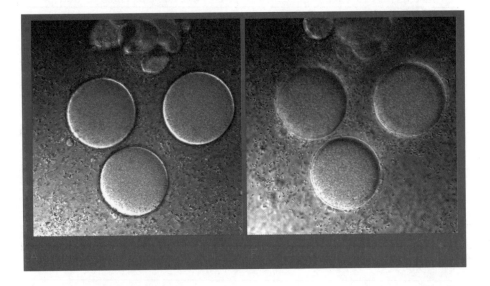

Figure 64. *Bright field images of panels A and F demonstrate the physical integrity of the eggs over the course of the experiment.*

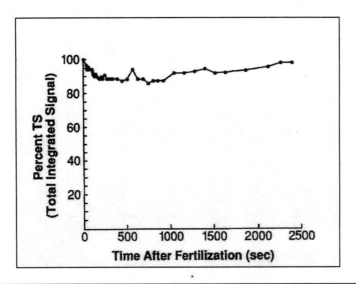

Figure 65. *Stability of the total fluorescent signal over the time-course experiments. No changes of the intensity are apparent.*

In contrast, a graph of %Z versus time following fertilization (fig. 66) showed a close fit to an exponential function (fig. 67; $t_{1/2}$=23.3 minutes, r=0.92).

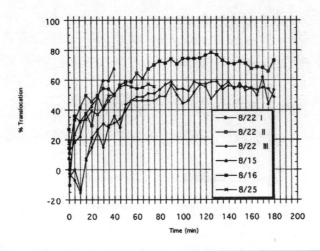

Figure 66. *Graph of %Z as a function of observation time for six* L. Pictus *eggs subsequent to fertilization.*

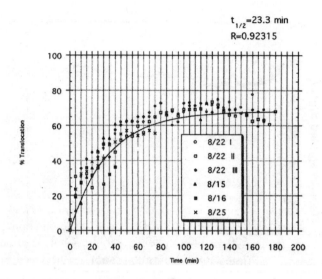

Figure 67. *The data were normalized to zero for the starting point of the translocation in order to analyze the process kinetics. An exponential function with a $t_{1/2}$ of approximately 23 minutes yielded a correlation coefficient of 0.92.*

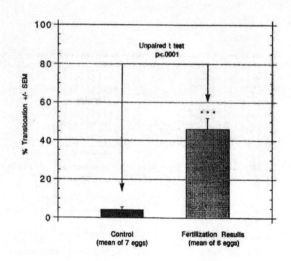

Figure 68. *Statistical analysis of translocation time-course experiments. %Z at 40 minutes had a value of approximately 50%.*

The %Z at 40 minutes subsequent to fertilization was compared to the %Z in unfertilized control eggs (fig. 68; 46.14±4.9 vs. 4.23±1.42 respectively, N=7, p<0.0001 unpaired t test).

Some eggs were followed through up to the second cell cycle and manifested a general decrease in fluorescent signal. The fact that loaded cells went on to divide and produce apparently viable embryos suggests that the phorbol ester probe was not toxic at the concentrations employed here.

In vitro subcellular fractionation experiments were conducted to determine if a translocation of PKC subsequent to fertilization could be detected using an independent method. The results of both ^{3}H-PDBU and N-terminal antibody dot blot experiments were similar (tab. 10). Both techniques demonstrated either a trend or a statistically significant net movement of radioactive phorbol ester label or immunoreactivity from the cytosolic fraction to the membrane fraction in eggs incubated with sperm for 40 minutes when compared to eggs incubated either with boiled sperm or control eggs. A two-way ANOVA followed by post-hoc hypothesis tests showed that the ratio of cytosolic to membrane fraction (marked with an asterisk in tab. 10) was significantly decreased (F=4.065 p< 0.038).

The primary finding of the experiments presented in this chapter was a sustained reorganization and subsequent translocation of the fluorescent phorbol ester probe to the vicinity of the cortical granules and plasma membrane after fertilization in sea urchin eggs. The phenomenon was statistically significant and uniformly reproducible. The increase in %Z reached half maximal levels approximately 20 minutes post-fertilization which corresponds to a point midway through the first cell cycle (Steinhardt, 1990b; Steinhardt, 1990a; Hinegardner, 1975). It is possible to interpret these results as a change in the distribution of the label as a result of the movement of PKC from the cytosol to the vicinity of the plasma membrane because: (1) the fluorescent phorbol esters bind to purified PKC with high affinity (Kd < 10 nM); (2) the concentration of fluorescent phorbol ester was not sufficient to activate PKC; and (3) ^{3}H-PDBU binding and dot blots using the polyclonal antibody against sea urchin PKC manifested an increase in both label and immunoreactivity associated with the membrane fraction 40 minutes after fertilization when compared with control eggs.

While a specific role for this fertilization-induced translocation of fluorescently-labeled phorbol ester remains to be elucidated, it is reasonable to hypothesize that it represents some dynamic change in the activation of PKC, similar to an analogous process occurring in mammalian CNSs after associative conditioning (Olds et al., 1989).

Condition		ROD[1]	DPM[2]
Control:	Cytosol	0.40 ± 0.08	22,769 ± 1907
	Membrane	0.22 ± 0.01	22,806 ± 1543
RATIO (Cytosol/Membrane)		**1.82**	**1.00**
Boiled Sperm:	Cytosol	0.71 ± 0.03	22,557 ± 2509
	Membrane	0.39 ± 0.02	25,313 ± 1948
RATIO (Cytosol/Membrane)		**1.82**	**0.89**
Fertilized:	Cytosol	0.32 ± 0.02[*]	16,797 ± 2243
	Membrane	0.22 ± 0.01[*]	23,955 ± 2225
RATIO (Cytosol/Membrane)		**1.45**	**0.70**

[*] A two-way MANOVA showed a significant interaction between Condition (Control, Boiled Sperm Control, Fertilized) and Fraction (Cytosol or Membrane) $p < 0.037$ $F = 4.065$.

1. Relative Optical Density reading from an M1 Imaging System.
2. Disintigrations per minute.

Table 10. *Confirmation experiments:* in vitro *subcellular fractionation of sea urchin egg PKC in S. purpuratus.*

Thus, it might be that phosphorylation of serine and threonine amino acid residues on PKC protein substrates may play a crucial role in the events subsequent to fertilization. One such possible substrate is the Na^+/K^+ antiporter (Eckberg and Palazzo, 1992; Bloom et al., 1995). However, the time course of the label translocation far outlasts the kinetics of transporter activation (Eckberg and Carroll, 1995; Eckberg et al., 1987; Eckberg and Szuts, 1993). An additional substrate may be the G-protein, dynamin, which has been shown to play a major role in the endocytic pathway of secretory vesicles (Robinson et al., 1994). Finally, the presence of cp20 in *S. purpuratus* eggs was recently demonstrated (Nelson et al. 1996a). Cp20 could be a robust candidate as a PKC substrate with physiological signaling effects, constituting a direct molecular parallelism between developmental biochemistry and synaptic plasticity mechanisms. A schematic diagram of how fertilization-induced PKC activation might affect the first cell cycle is shown in fig. 69.

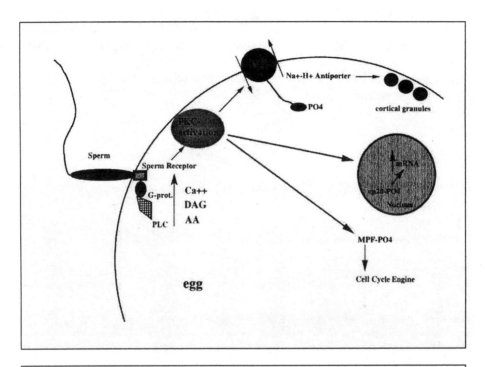

Figure 69. *A schematic diagram illustrating several putative roles for PKC in the events subsequent to fertilization of the sea urchin egg. The sperm interacts with a G-protein-linked receptor to activate phospholipase C and phospholipase A_2, resulting in an increase in diacylglycerol and arachidonic acids (AA) along with an IP_3- and ryanodine-receptor dependent mobilization of Ca^{2+}. These factors act synergistically (perhaps with AA from the cortical granules) to maximally activate protein kinase C. Activation of PKC results in phosphorylation of serine and/or threonine residues on the Na^+-H^+ antiporter, on cp20 and on other small GTP-binding proteins, resulting in a coordinated progression out of G2 arrest and into mitosis.*

The present *in vivo* translocation of NBD-phorbol probe to the vicinity of the egg plasma membrane subsequent to fertilization is likely to represent a reorganization of protein kinase C localization. The use of a ratio of fluorescence associated with the membrane compared (MS) to the total signal (TS) for the egg allowed a quantification of the translocation that was independent of between-egg differences in the amount of quantum fluorescence yield of the NBD probe. The present results extend other's observations by demonstrating that the operational *membrane-association* of PKC in the test-tube represents an *in vivo* association of the enzyme with the area of the cortical granules and plasma membrane. This localization of the activation may be important in light of the very high content of arachidonic acid in the cortical granule membranes (Detering et al., 1977). Arachidonic acid is a potent activator of PKC, acting in synergy with diacylglycerol (see Lester, 1992). Thus, PKC may, via its translocation to this milieu, undergo a sustained activation, qualitatively similar to the constitutive activation seen in other contexts such as learning and memory (Olds et al., 1989).

CONCLUSIONS

This thesis describes the purification, cloning, biochemical and spectroscopical characterization of cp20. After the aminoacidic sequence was found, a complete study of the secondary structure of cp20 was achieved. Cp20 has proven to be a calcium-binding protein with potential cellular signaling significance. In particular, a remarkable conformational switch upon calcium binding was characterized by CD, fluorescence spectroscopy, reverse phase chromatography and non-denaturating gel electrophoresis. In the passage from the apo-form to the calcium-bound conformation, cp20 becomes more globular and compact, but the overall polarity does not mutate.

Since the main cellular role of cp20 is the inhibition of calcium-dependent potassium channels, which causes a reduction of the ionic current and a hyperpolarisation of the neuron, a question opened by the present study is to which extent the calcium-induced conformational change has to do with the translocation of cp20 into the membrane. It is known that phosphorylated cp20 is preferentially found in the membrane, while the dephosphorylated protein is in the cytosol, and it is also known that cp20 is phosphorylated as a trigger for potassium channel inhibition (see also McPhie, 1994). Preliminary results from electrophysiological experiments indicate that the presence of Ca^{2+} in the vicinity of the membrane is indeed necessary for cp20 to be active towards ionic channels, and the concentration needed is similar to that for cp20 to change secondary structure (P. Gusev and D.L. Alkon, NIH, Bethesda, MD, USA, personal communication).

Is there a relationship between non-covalent calcium binding and covalent phosphorylation? PKC itself is activated by calcium (at an even higher concentration than that causing the cp20 transitional switch), so it might be speculated that calcium binding represents a first step of temporary activation (in that it is a reversible equilibrium), while if PKC becomes activated, cp20 undergoes a more stable covalent activation. It is still unclear whether or not the phosphorylation is the direct signal for membrane translocation. Ca^{2+} is not a candidate for this role, because the polarity of the protein is not affected by calcium binding. Generally proteins are anchored to the intracellular side of the cytosolic membrane by being covalently attached to a fatty acid chain. Cp20 has three possible post-translational modification sites of this kind (two myristoylations and one isoprenoylation). Ca^{2+} binding and/or phosphorylation could be signals to (1) trigger another post-translational modification driving cp20 to the membrane and (2) interact with potassium channels.

An interesting EF-hand calcium-binding protein, called Recoverin, was recently discovered and characterized (Ames et al., 1994 and 1995b). Recoverin is a specific protein of the retina, and it is covalently modified my a myristoyl group. Under resting conditions, the fatty chain of myristoate is bound to a hydrophobic binding site inside the protein, and cannot therefore interact with the membrane.

When Recoverin binds calcium, a conformational change is triggered which extrudes the myristoyl moiety outside of the protein, and Recoverin translocates to the membrane. This mechanism is likely to be quite general, and could act as one of the modulatory processes of cp20. Alternatively, Ca^{2+} binding and phosphorylation might be similarly related.

From the results presented here, it is possible to propose a simple and schematic model of cp20 activation. The details cannot be hypothesized at this stage, but more experiments are in progress, as discussed below. In resting conditions, cp20 only binds one mole of calcium, and is inactive; the neuron is normally polarized and the neural pathway is at ground state. When the neuron receives strong enough signals through the dendritic tree, it reaches the action potential, shooting an axonic impulse; at this point the intracellular calcium concentration temporarily increases to a level at which (1) calcium dependent potassium channels are activated (the neuron starts repolarizing), and (2) cp20 binds a second mole of Ca^{2+}, therefore switching secondary structure to an "active state". In this conformation, cp20 can inhibit potassium channels, and positively feed back neural activity. When the cell stops shooting, the calcium concentration decreases back to ground level, cp20 is deactivated, and calcium dependent channels close their gates. In contrast, if a particularly strong train of stimuli hits the neuron (such as in case of tetanic stimulation and LTP, or in case of non-Hebbian associative conditioning), $[Ca^{2+}]$ increases to a much higher extent, and remains at an excited level for a longer time. In these conditions, not only calcium-dependent potassium channels are open, and cp20 is active, but PKC becomes also stimulated. PKC phosphorylates cp20, and even when the calcium level starts decreasing, cp20 remains in an active state, i.e. it keeps inhibiting calcium dependent potassium channels disregarding the Ca^{2+} concentration. This mechanism could explain the neuron's prolonged activity after cp20 microinjection, and proposes a direct link between protein activation and cellular physiology, which is in turn related to behavior, learning and memory. This schematic model is summarized in fig. 70.

This hypothetical model is supported by the structural studies presented in this thesis and by the preliminary finding that Ca^{2+} is needed at a similar concentration for inducing the secondary structure transition and for inhibiting potassium channels in electrophysiological experiments.

It should be noted that phosphorylation also is, on a different scale from calcium binding, a reversible equilibrium, because phosphatases can remove this post-translational modification from cp20 (Nelson and Alkon, 1995). However, on an action potential time-scale, the scheme of fig. 70 is sufficiently accurate.

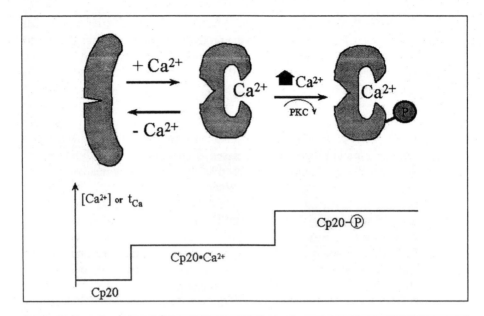

Figure 70. *Speculative mechanism of the activation pathway of cp20. At the ground state, cp20 is in an inactive, elongated shape. When [Ca^{2+}] increases, cp20 switches secondary structure to a more globular and compact shape. In this conformation, an interaction site (schematized by the crack on the left of the protein) becomes available to inhibit directly or indirectly calcium dependent potassium channels. If calcium concentration decreases, the equilibrium is reversed to the ground state. In contrast, if Ca^{2+} concentration further increases or remains high for a longer time, PKC becomes activated and phosphorylates cp20. This covalent modification leaves cp20 in an active state for long time even at lower calcium concentrations.*

The research line exposed in the chapters of this thesis must be continued and expanded to reveal more completely the biochemical pathway involving cp20, PKC and calcium-dependent potassium channels. Future perspectives include:

a) A deeper characterization of cp20's structure, particularly with respect to tertiary structure and dynamic behavior (by high-frequency ^{13}C and ^{15}N nuclear magnetic resonance).

b) The study of cp20 aggregation phenomenon (by CD in transparent detergents).

c) The confirmation of phosphorylated cp20's secondary structure and its variation with calcium binding (by CD and FT-IR).

d) A detailed characterization of the influence of Ca^{2+} concentration on cp20's inhibition of potassium channels (by patch clamp).

e) The conformational analysis of cp20 ligand binding, with particular attention to GTP (by low-energy CD)

f) The investigation of myristoylation and the other potential post-translational modifications (by biochemical and spectroscopical methods).

g) Imaging both PKC and cp20 in living neurons and astrocytes upon cellular activation (by laser confocal microscopy).

In conclusion, the present thesis brought a contribution to the study of the molecular mechanisms of synaptic plasticity. Several questions regarding the characterization of cp20 have been answered, and many more have been raised. Despite the ongoing exploration, the structure-activity relationship for cp20 is closer to being fully understood.

APPENDIX

A. Coordinate systems for dendritic spines.

B. Ligand binding study of HSA by CD.

C. Post-translational modifications of B-50 by MS.

A few research lines were developed in parallel with the main project of this thesis. Some of them were independent efforts, such as the definition of a new coordinate system for the dendritic trees of neurons (part A), others were preliminary to the work developed in the thesis, such as the set up of the CD analysis of proteins carried out on human serum albumin (part B). Some others, such as the preliminary structural study of the neuronal proteins B-50 and neurogranin, began to apply the techniques developed with the study of cp20 to other substrates of synaptic plasticity (part C) . The research described in this appendix is not directly related to the main stream of this thesis, but was conducted during the same time frame and addresses the same issue of rationalizing the relationship between physiological activity, cellular architecture and molecular structure.

The discovery and the study of cp20 made it possible to build intracellular and intercellular models of the neuronal mechanisms of synaptic plasticity (for a review, Alkon, 1989). Of fundamental importance in describing a neuronís activity and constructing biologically plausible neural networks is the unambiguous description of its smallest element of input in the integration process. Among neuronal input units are the synaptic spines, highly regulated and coordinated elements on the dendrites, exchanging both electrical signals and molecules with the soma along the dendritic branches. Mapping the physiological parameters of dendritic branches and their spines in anatomically compatible coordinates is important because of the interactions between "close" spines and between spines and the soma. In part A of this appendix, a simple method is presented for quantitatively locating dendritic spines by separating their coordinates into two components. The first takes into account the position of the dendritic branch on which the spine lies. In this component, the distance between a branch and the soma is given by the number of bifurcations along the dendrite ("level"). The difference in this parameter between any two spines ("distance") was formulaically described in terms of the level of the common bifurcation farthest from the soma ("generator"). The second component of a spine's location is its position on the dendritic branch. This system is fully analytical and easily implementable. It also defines a biologically plausible distance between any two spines , and between a spine and the soma. Based on this labeling method, a coordinate system is presented in which a spine is described by a matrix encoding physiological parameters of the generating branches. A second set of coordinates is introduced to describe a neural state with a matrix of spine parameters. Finally, a third matrix notation is proposed to take into account interactions between spines. This treatment leads to some interesting speculations, such as the possibility of describing input dynamics of a neuron in terms of operators on vector spaces.

Part B describes a more experimental and less speculative research line, which was used to set up the CD analysis method for protein secondary structure determination. The commercial protein human serum albumin represented a suitable model, because its physiological activity, ligand binding, can be also detected by CD. Several post-translational modifications (some of which are known to occur under physiological conditions) were chemically induced on HSA, and the secondary structure was measured in the native protein and in modified samples. Binding properties were also studied on native and modified HSA by low-energy CD, and the correlation between secondary structure and biochemical activity was determined. The use of CD to study protein ligand binding can be also applied to the characterization of cp20. The applicability of IS-MS to crude partially purified HSA was also investigated.

In part C, the study of another important neuronal substrate for synaptic plasticity, B-50, is reported. B-50 is in some extent similar to cp20, in that it is a small MW (25 kD) proteins which is anchored to the cytosolic membrane through a covalently attached palmitoyl group. It binds Calmodulin, but this activity is inhibited by the phosphorylation of a single aminoacidic residue by PKC. Other potential covalent modifications include an ADP-rybosylation. Therefore the post-translational modifications of B-50 represent an important target for IS-MS analysis. A partial purification protocol based on a simple acidic extraction was developed to obtain a suitable form of protein for on-line HPLC/IS-MS measurements. Another neural substrate of PKC, neurogranin, was also detected in the protein extract from μg samples. After presenting the preliminary findings of this research, the perspective of future projects are outlined.

A.   Coordinate systems for dendritic spines: a somatocentric approach.

It is widely accepted that the neuron is not a simple switch or unit of transmission, but rather a complex system able to handle a large amount of information (Churchland and Seynowski, 1992, pages 37-44). The integration of a large set of inputs by a single neuron constitutes a process of complex computation, ultimately responsible for many if not all macroscopic abilities of nervous systems (fig. 71).

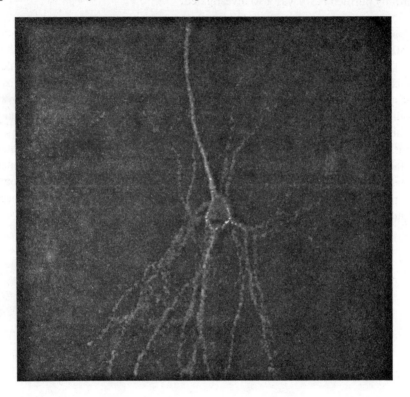

Figure 71. *Confocal microscopy of a pyramidal cell from the CA1 region of rat hippocampus. Cultured neurons were stained with lucifer yellow and scanned with an argon-ion laser. The soma is visible in the middle, several dendritic trees at the bottom, and the axon at the top of the figure (this picture is a courtesy of C. Collin and M. Segal, NIH, Bethesda, MD, USA).*

It has been proposed that even higher cognitive functions in mammals, such as learning and memory, find their molecular correlates not only in cellular interaction among neurons, but also in intraneuronal events involving subcellular organelles, the dendritic spines (fig. 72; for a complete discussion, see e.g. Moser et al., 1994, and references therein).

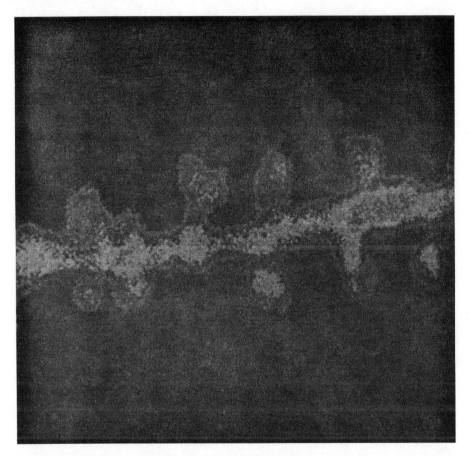

Figure 72. *100-fold enlargement of a dendritic branch from fig. 71, several spines are evident (courtesy of C. Collin and M. Segal).*

Spines, small protuberances, about 20,000 to a cortical pyramidal cell, are one type of synaptic loci for the input in the neuron-to-neuron interactions, representing therefore a unit of integration. The neuron can also receive stimuli through direct synaptic contacts on the soma, although dendritic connections seem to play a pivotal role in the determination and regulation of fine-tuned processes of the neuron, such as association (fig. 73; for a review on spine synaptic inputs, see Harris and Kater, 1994).

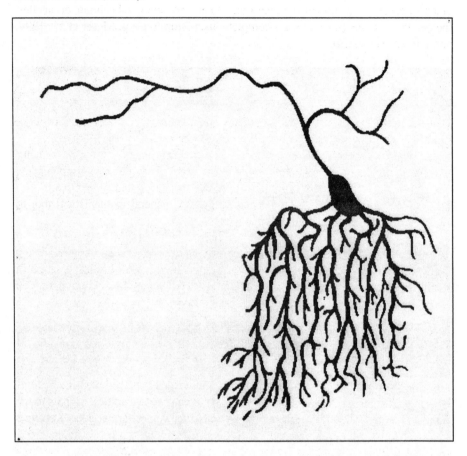

Figure 73. *Schematic draw of a neuron such as in fig. 71. The anatomy of the dendritic tree is shown.*

In addition, not all the synapses on the dendritic branches involve spines, and in general the precise functional role of dendritic spines is still under debate (Segal, 1995; Collin et al., 1996).

It has been postulated that, in principle, a complete map of synaptic activities would determine the state of the neuron entirely (Segev et al., 1995), and it appears evident that while the contribution of direct connections on the soma is more easily computable, dendritic inputs are both crucial and difficult to model. Even though the physiology and the biochemistry of dendritic spines are far from well known, a biologically meaningful way of describing spine location with respect to the soma or to other spines is clearly needed.

Furthermore, in classical neural network models, any intracellular cytoarchitecture and compartmentalization are ignored, and the system is completely specified when the connection weight between any two neurons is known (Dayhoff, 1990). Based on the neurobiological research, this approach seems to be too crude and simple to achieve the computational power of natural nervous systems, and indeed several models have been proposed in which the neuron is itself a neural network (Anderson and Sutton, 1995). Therefore, of fundamental importance for those neural network researchers who wish to design biologically plausible models, is a system which numerically labels dendritic spines in a manner suitable to take into account physiological parameters. Moreover, there is no need to topologically distinguish dendritic spines from other dendritic synapses, thereafter a coordinate system for spines or dendritic branches furnishes a general mathematical tool to model highly modulatory neuronal inputs.

In order to match these needs, suitable coordinates for dendritic spines should fulfill the following characteristics, which are discussed below: (1) the position of a spine in these coordinates must encode (or be incorporated in) a simple definition of a distance between the spine itself and the soma of the neuron; (2) a distance between spines along the dendrites must be defined based on their positions relative to the soma; (3) the mathematics developed with the definition(s) of distance(s) must be easily implementable.

The first characteristic is justified by the need to describe the complex interactions between the spines and the soma. For instance, spines send electrical signals to the soma and as a first approximation one may consider spines farther away to have a smaller weight in the final integration of the neuron (Gardner, 1993). On the other hand, small molecules and biopolymers travel along the dendrites in both directions, and both active transportation and passive diffusion seem to play a role in this process (Gorenstein et al., 1985, and references therein). An important example is given by calcium ion, a second messenger that also plays a role in dendritic action potential (Muller et al., 1991). In the case of the soma delivering molecules to the spines, it is likely that a smaller amount of substance will be available for spines farther from the soma (Dudai, 1989, pages 134-138). A particular example of this process will be discussed in the section "Applications" of this appendix. It is interesting to note that in order to describe the two above mentioned examples (electric signaling and molecular transportation) different information content (length and size of the branches, number of bifurcations and so on) may in principle be useful in the definition of distance of the spine from the soma. In general it is convenient to develop a flexible coordinate system that allows the definition of several distances, or of one distance with several possible biological applications.

The second characteristic is necessary in order to describe interactions between spines: evidence suggests that adjacent spines undergo mechanisms of cooperativity or mutual inhibition as a subcellular basis of associative learning and memory (Dayhoff et al., 1994; Alkon, 1995). Models have been proposed in which the association between two distinct neural pathways, for instance in the case of classical (Pavlovian) conditioning, is explainable in terms of association between spines in single neurons shared by the two pathways (Hosokawa et al., 1995). In this model, close spines are simply likely to interact more efficiently. A more elaborate model holds that each cluster of spines (on a branch or even a group of branches) can also interact with other close clusters, thus forming the second of several levels of association, eventually integrated by the neuron (Alkon et al., 1994). In these models, the definition of a distance between spines would allow the description of a "sphere" of interaction around a spine.

The two above discussed characteristics imply that the anatomy (topology) of the neuron should be conserved (or taken into account) in the coordinate system: for example, locating spines in standard Cartesian coordinates is a very inefficient system to describe processes along the dendrites, and any correction factors make the system inconveniently complicated. Nonetheless, there are some neural phenomena that are actually well described in simple Cartesian coordinates, such as the possibility for spatially close spines to receive synaptic contacts from the same axon (Harris, 1995), or the influence of the concentration of specific extracellular substances on intracellular activities. The final description of the neural state will have to consider both the Cartesian and the non-Cartesian coordinate contribution, although only the second component is the domain of this appendix.

Method

The numerical system present here assigns a unique pair of numbers to each spine, designating its position on the dendritic tree. The first number indicates on which branch of the dendrite the spine is located. The second measures how far along this dendritic branch the spine lies (the spatial distance from the last bifurcation of the dendrite). A *topological distance* between any spine and the soma is defined by the number of branch bifurcations between the branch on which the spine is located and the soma itself, *and* by how physically close the spine is to the bifurcation which defines the branch. This easily generalizes to a distance between any two spines. The first number in the pair is useful to describe those interactions between the soma and a specific spine, or between two spines, which depend more acutely on the number of bifurcations along the dendritic branches than they depend on the actual measured distance of the spine from the soma. The use of the second number, however, targets a specific spine after finding the branch on which it lies, like a house number on a named street.

Of primary importance, then, is finding a simple system to number the dendritic branches themselves. Directions from the soma to a specific dendritic branch may be specified by a series of choices, either "right" or "left" at each bifurcation of the branch. Right and left are arbitrarily defined in three-dimensional space using a choice of orientation at each bifurcation. It is opportune to standardize this choice to identify univocally a given branch by its label. An example of such a standardization is briefly described here: first orient the neuron in three-dimensional space with a choice of orthogonal axes (labeled x, y and z.) centered at the soma. This choice of Euclidean coordinates is the only part of the orientation which is arbitrarily assigned. At each bifurcation, the two protruding branches are distinguished from one another by any plane which separates them in three-space. Translate the coordinate axes to the point of bifurcation. If the two branches span a plane which is not perpendicular to the xy-plane, they can be separated by a plane through the z-axis as follows. If the yz-plane separates the two branches, call "right" the branch in the positive x-hemispace and "left" the branch in the negative x-hemispace. If the plane does not separate them, simply rotate the coordinate system about the z-axis counterclockwise until it does; label "right" and "left" as prescribed. If the two branches lie in a plane which is perpendicular to the xy-plane, separate them by the xy-plane after a rotation of the coordinate system counterclockwise around the x *axis*. Label the branch in the positive z-axis "right" and the branch in the negative z-axis "left".

It should be noted that there are theoretically no circumstances in which there is a three-way (or more) bifurcation. One may consider such occurrences as single bifurcations followed closely by secondary bifurcations. Since there is only one path from the soma to any branch within a specific dendritic tree, the address of the branch is uniquely specified by a sequence of "left"s and "right"s, indicating directions from the soma to the branch. This awkward language is replaced with binary code; each dendritic branch is labeled by a unique binary number, where 1 stands for "left" and 0 stands for "right". This number is converted to the decimal system to obtain a unique positive integer assigned to each branch. The branch protruding from the soma which defines the dendritic tree is labeled 1.

With this numbering system in place, one can easily calculate the "branch-distance" of the spine to the soma, defined as the number of bifurcations there are in between the soma and the branch. This distance will be called the *level* of the branch (fig. 74).

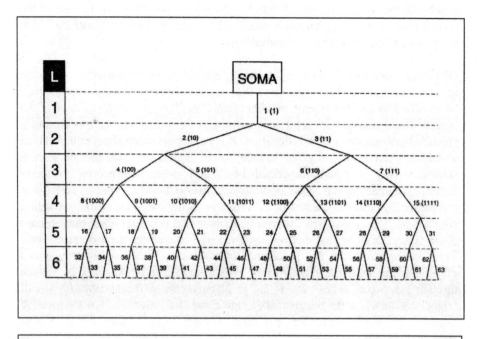

Figure 74. *Schematization of the numbering system for a dendritic tree. The branch level is reported on the left. The neuron is oriented as in figures 71 and 73.*

If we refer to the binary number associated with this branch, it is simply the number of 0's or 1's in the number; each 0 or 1 indicates one bifurcation. The equivalent description in decimal notation is to define the level l of a branch x to be

$$l(x) = 1+Int(\log_2 x)$$

where $Int()$ is the integer of $()$.

The topological distance between a spine and the soma is then defined by a pair of numbers; the first is the level of the branch, and the second is the measured distance between the spine and the bifurcation generating the branch. The ordering amongst the spines is the following; any spine with a lower level is closer to the soma than any spine with a higher level. Amongst spines which are at an equal level, those which have a smaller second number in their addresses are closer to the soma than those with a larger one.

This framework suggests a notion of distance between two spines as well. Such a distance must be faithful to the anatomical path taken in traveling from one spine to another. The shortest path from one spine to another is the (unique) route which goes from the first spine "up" the dendritic tree to the first branch it has in common with the other spine, then "down" different branches until it reaches the second spine. It is therefore useful to define the *generator* of two spines, or the branch farthest from the soma which is in the path from the soma to *both* spines (fig. 74). A *parent* of a spine is defined to be any branch in the path from the soma to the spine. The generator of two spines is thus the common parent which is farthest from the soma. The topological distance between two spines has two components, defined as follows: the first is the sum of the differences in levels between each spine and the generator. The second component is the sum of the second numbers of the addresses of the spines. Two spines are closer to each other than two other spines if either the first component of the distance between them is smaller than that of the other pair of spines, or if the first component is the same and the second is smaller. To calculate the first component of the distance between two spines, we need only know on which branches they lie. Let x and y be the positions of two branches. Then the distance between the branches (which is also the first component in the pair which defines the topological distance between two spines) is described by:

$$d(x,y) = l(x) + l(y) - 2l(g(x,y)) \qquad (*)$$

where $g(x,y)$ is the position of the generator of x and y. Notice that the generator of two spines depends only on the branches on which the spines lie.

It must be feasible to calculate the level of the generator of two spines on branches labeled x and y. Assume for the moment that the two spines in question have the same level. If x and y were the same number, the level of the generator would be simply that of x (or y). If they are not the same number, we want to find the unique parent branch one level lower ("higher" on the dendritic tree) of each spine and compare these two. If they are the same, the generator is found, and if not, this process is continued. The corresponding mathematics to specifying the level of the parent of the branch x one level lower is dividing by 2 (using decimal notation), and taking the integer of $x/2$. This is an iterative process: at each step, we evaluate $Int(2^{-n}x)$ and $Int(2^{-n}y)$ and compare them, where n is the number of times the iteration has occurred (n begins at 0, which compares the numbers x and y themselves). The level of the generator will be the level of the number $Int(2^{-n}x)$ when it first coincides with $Int(2^{-n}y)$, i.e. for minimal n.

To formalize this process mathematically, an expression should be found which is 1 when we have a common parent, and 0 otherwise. The level of the generator is then a sum of these numbers, for there are precisely $l(g(x,y))$ common parent branches of x and y. It is sufficient to find such an expression which is between 0 and 1 for each n, and strictly less than 1 if and only if the $n\underline{th}$ iteration of moving up along the tree (decreasing the level by one) does not yield a common parent; applying the $Int()$ function makes the expression equal to 1 when we have a common parent, and 0 otherwise, as desired. The expression

$$[Int(2^{-n}x)]^{-1} - [Int(2^{-n}y)]^{-1}|$$

has these properties, except it equals 0 if and only if the two numbers are equal. The expression:

$$\cos\{(\pi/2)|[Int(2^{-n}x)]^{-1} - [Int(2^{-n}y)]^{-1}|\}$$

equals 1 exactly when we have a common parent, and is otherwise strictly less than 1. Apply $Int()$, and we have a new expression which is 1 when we have found common parents of x and y and 0 otherwise. Summing over n, the level of the generator is obtained:

$$l(g(x,y)) = \Sigma Int(\cos\{(\pi/2)|[Int(2^{-n}x)]^{-1} - [Int(2^{-n}y)]^{-1}|\})$$

where the sum is from $n=0$ to $n=l(x)$.

Now for the more general case. Without loss of generality, assume that $l(x) \bullet l(y)$. The unique parent of y is found which is on the same level of x , and then apply the method above. Call this parent y'. We have:

$$y' = Int(2^{-(l(y)-l(x))}y).$$

Any easy check shows that $l(y')=l(x)$. The formula given in (*) is now calculable.

This method of numbering branches and spines establishes two closely related notions of somatocentric coordinates. The first ("branch coordinates") describes a particular spine, where the $i\underline{th}$ coordinate corresponds to the $i\underline{th}$ branch in the path from the soma to the spine. Alternatively, the "spine coordinates" describe a dendritic tree, where the $i\underline{th}$ coordinate corresponds to the $i\underline{th}$ spine in the tree.

Applications

There are several applications of this method, three of which we briefly attempt to explore here. The first is to model a single spine, describing several physiological features which can be represented through the coordinate system developed in this appendix. The second is a model for the entire neural input map, described in terms of dendritic spine parameters. The third application is to model the interactions between any two spines of the neuron, such as association, cooperativity or mutual inhibition.

The first application of the somatocentric coordinates constructed in this system is a matrix description of a single spine. Branch coordinates allow an efficient tabulation of the information about a spine, including as many or as few parameters as desired, in a matrix of real numbers or functions. For the purposes of this exposition, we consider a matrix as a collection of column vectors, each of which will represent some category of information. The coordinate system for the matrices is that described in section Method; the $i\underline{th}$ entry in each vector corresponds to the $i\underline{th}$ branch in the path from the soma to the spine in question. Thus if there are n branches in this path (including the branch on which the spine lies), there will be n entries in each vector. With this coordinate system in place, the first vector in the matrix associated to a specific spine is the *branch vector*, which consists of the sequence of 0's and 1's corresponding to the path from the soma to the spine. The branch vector is obviously motivation for the choice of coordinate system, but the coordinate system allows us to incorporate far more information in vectors of the same form (fig. 75). For example, a vector of branch length can be defined, where the $i\underline{th}$ entry is the length of the $i\underline{th}$ branch; the last entry will be weighted in the sense that it will record the length of the last branch from the defining bifurcation to the spine, not the entire length of the branch.

Correspondingly, the branch diameter vector can be defined; in this case, the last entry will not need to be weighted because of the position of the spine on the branch. In fact, one can define a column vector for any local physiological property, i.e. any property that does not depend on the state of the entire neuron, rather is well-defined for individual branches. In particular information pertaining to the electrical properties of the branch, such as conductivity, permeability to several specific ions, presence of active pumps etc. can be neatly tabulated. Such information can be used to calculate the contribution of a spine's electrical activity to the soma potential. This method, in a simplified version, can also be applied to a description of non-spine mediated dendritic synapses.

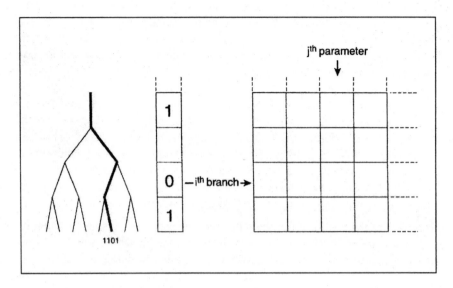

Figure 75. *"Branch" coordinates for spines; the resulting matrix describes a single spine in terms of its branch parameters.*

As an example of the utility of this description, a problem is discussed below concerning molecular transportation. When a specific dendritic spine drastically changes its activity, for instance following a sustained activation of the neural pathway in which it is involved, it undergoes functional and morphological changes that require the synthesis of new proteins and thereafter the activation of specific translational and/or transcriptional processes (Alkon et al., 1990; Nelson et al., 1991). An extremely intriguing open problem in neurobiology is: how can the soma (where the genes are expressed) specifically target the activated spine (where the proteins carry out their function) with the delivery of a macromolecule? A model has recently been proposed in which a regulatory protein exists both within the nucleus, associated with plasticity-related genes, and in the dendrites (Olds, 1995), associated with the spine-localized polyribosomes (Steward and Falk, 1986). Instead of binding upstream to the exon, such transcription factor would bind, when activated, within the c-strand of the exon itself in order to enhance or promote gene transcription (fig. 76). The resulting mRNA would have the desirable effect of also being a "receptor" for such protein.

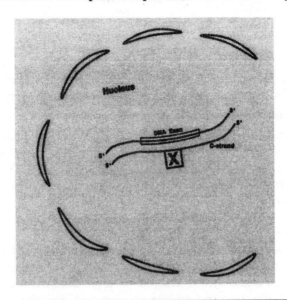

Figure 76. *The "zip-coding hypothesis": the regulatory protein binds to the c-strand of the exon, behaving as an anomalous transcription activator (this picture is a courtesy of J.L. Olds, GMU, Fairfax, VA, USA).*

The mRNA product would then be carried along the whole dendritic tree by means of the active transport machinery (Huber et al., 1993; Tanaka et al., 1994), but would be selectively "captured" by those spines containing the activated promoter itself (fig. 77).

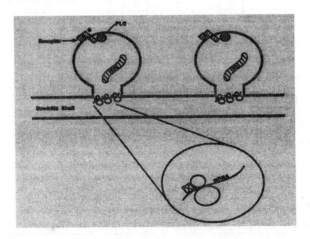

Figure 77. *The dendritic phase of the "zip-coding hypothesis": only spines in which the promoter is activated can receive the mRNA and start the protein expression process (courtesy of J.L. Olds).*

How much of the macromolecule produced in the soma actually reaches the targeted spine? A simple estimate of the quantity of gene available for a specific spine is obtainable by its level as defined in the Method; at every bifurcation the amount of the macromolecule is roughly halved, so we might expect that $1/2^n$ of the material reaches a spine on the $n\underline{th}$ branch. A more accurate description is provided by considering other parameters such as the diameter of each branch (at each bifurcation, more material flows into larger branches) or the activity of the transportation machinery in each branch, and could be computed by means of the spine matrix in branch coordinates introduced above.

A more general application of our coordinate system might be used to handle information concerning the whole neuron, whose electrical input basically depends on the activity of its dendritic trees. The same idea used for a matrix associated to a single spine can be employed to describe an entire dendritic tree. This matrix is also written in somatocentric coordinates, but now the spine coordinates are considered instead of the branch coordinates; the $i\underline{th}$ entry of the tree matrix corresponds to the $i\underline{th}$ spine (fig. 78).

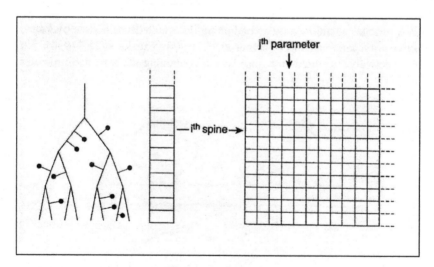

Figure 78. *"Spine" coordinates for dendritic trees; the resulting matrix describes an entire dendritic tree in terms of its spine parameters.*

As described for the matrix associated to a spine, the dendritic tree matrix may include any parameters which are well-defined for individual spines. Each column in the matrix will be a vector of a chosen parameter such as spine activity, and the

$i\underline{th}$ entry of this vector will be the activity of the $i\underline{th}$ spine in the tree. Whereas a spine matrix records data associated to a particular spine and its relationship to the soma, the dendritic tree matrix tabulates information about every spine, such as each spine's weight, activity, etc., without specifying anything about the branches which lead from the soma. Interestingly, some of the information needed for the dendritic tree matrix, i.e. to model the neuron, might be obtained from the spine matrix.

An important feature characterizing spines is their relationship with each other. Spine interaction heavily influences neural activity, and models have been proposed in which the non-linear summation of the electrical contribution of specific spines on a dendritic tree constitutes the subcellular basis of associative memory such as in classical conditioning (Wang and Ross, 1992). The third example of applications of the coordinate system presented in this appendix is then a way of tabulating the interaction between any two spines on a dendritic tree. For each pair of spines, we would like to describe parameters of interaction, such as cooperation, inhibition, topological or electrical distance between the two spines, etc. For each parameter, we construct a matrix of the interactions of the spines; the $ij\underline{th}$ entry of the matrix is the parameter applied to the spines i and j (such as the topological distance between spines i and j). Placing these matrices together in a three-dimensional matrix, we have constructed a table of interactions between spines (fig. 79); the $ijk\underline{th}$ entry of the matrix is the $k\underline{th}$ parameter of spines i and j.

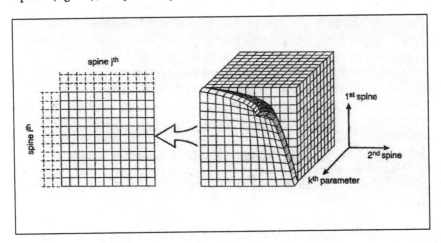

Figure 79. *Three-dimensional association matrix, built to describe the parameters of interaction between any two spines in a dendritic tree.*

Based on the information encoded in this matrix, correction factors might be introduced in both previous matrices (for spines and for the whole dendritic tree), for any parameter influenced by spine-to-spine interactions.

Discussion

A numerical algorithm useful for a mathematical description of dendritic branches, spines and their interactions with the soma and with each other was introduced in the present appendix. Two new coordinate systems, called *somatocentric*, where consequently introduced, in which the cellular anatomy of the neuron is conserved, in that only the space along the dendritic tree is considered. This numerical algorithm and these coordinate systems can therefore be adopted for every problem in which the structure of the dendrites is important. Although modern neurobiology suggests that this is indeed the vast majority of cases, it should be noted that other coordinates (Cartesian or polar for instance) may be used to describe phenomena that do not depend on the dendritic structure.

It is interesting to notice that the coordinate system described here, because it conserves the tree-like structure topology of dendrites, is actually a hypermetric space, i.e. it has an isosceles topology. This means that, given any three points a, b, c of the space (spines on the branches), their distances are such that $ab = bc \leq ac$, i.e. they form an isosceles triangle, with the basis shorter than the sides. This property can be formally proven, or can be intuitively verified on fig. 74. Hypermetric spaces are particularly interesting because, given a standard interaction potential (such as electrostatic, as in the dendritic spine case), which is inversely proportional to the distance, a system with n elements has 2^n energetic minima (for a review on this subject, Mezard et al., 1987). A peculiar characteristic of these systems (called *spin glasses*) is that they are *frustrated*, i.e. if a configuration is moved away from an energy minimum, it reaches equilibrium with a polynomial, rather than exponential, decay. This property was a key feature of a global neural network system proposed a few years ago to explain cerebral activity (Mezard et al., 1987), and it is intriguing to observe that the same mathematical formulation can be applied to a single element (the neuron) of that network.

The algorithm and coordinate systems proposed here are easily implementable and extremely flexible, as they can take into account as many parameters as needed. Some of these parameters might be experimentally measured and other may be calculated, which allows these coordinates to be used not only for a univocal description of an observed cellular system, but also as a basis for the design of biologically plausible neural networks, in which the subcellular input units (dendritic branches and/or spines) are also modeled. The flexibility of this system is mostly due to the assignment of an ordered integer number to each branch of the dendrite, and thereof to the possibility of building vectors whose elements correspond to specific branches or specific spines. This naturally leads to the use of matrix notation for physiological parameters. In particular, we suggested some applications that can be developed in the future, such as the detailed description of the properties of a dendritic spine, in terms of the parameters of the branches generating it, or the overall description of the properties of an entire dendritic tree, in terms of the parameters of the spines on it. Many processes can be described in these coordinates, such as electrical flow (which takes into account e.g. ion channel and active pump distribution, myelination, dissipation and path length), or molecular transportation (which depends on e.g. branch diameter, passive diffusion, active transportation and path length). Given an opportune definition of distance, spines can also be grouped in spheres or clusters, "close" in regard with a particular physiological property. Furthermore, interaction between spines or spine clusters can be expressed to model properties such as cooperativity or inhibition, responsible for non-linear behavior which is crucial for modeling associative memory.

Finally, the matrix notation adopted to describe physiological characteristics of spines and dendrites, introduces the possibility of using operators on vector spaces representing the dependence of considered parameters on time and neural activity. This structure suggests that the method presented in this appendix offers extremely powerful computational tools.

These somatocentric coordinate systems need to be tested on experimental data analysis and on existing or new neural networks to fully prove a physiological consistency and a computational utility, nonetheless they represent one of the first attempts to describe mathematically the subcellular structure of neuronal inputs with a direct anatomical approach.

B.   Ligand binding study of HSA by CD.

Human serum albumin (HSA) is the most abundant and important plasma protein, representing 60% of serum proteins (for reviews, see Peters, 1985; Kragh-Hansen, 1990 and Honorè, 1990). Although a broad range of physiological functions is reported for HSA, such as regulation of colloidosmotic pressure, buffer of the aminoacid concentrations and storage of divalent cations, it is widely accepted that the HSA's main role is ligand binding. Both exogenous (drugs and their metabolites) and endogenous molecules (fatty acids, hormones, aminoacids, ions and alkaloids) interact with HSA, and although HSA binding is generally considered a simple transport mechanism, it influences metabolism, distribution and elimination of most drugs.

On its highly globular polypeptidic chain of 585 aminoacids, HSA contains as many as 17 disulfide bridges, which fold the protein in a peculiar three-dimensional structure. Hydrophilic residues are exposed to the external surface, thus guaranteeing an extremely high solubility, while internal pockets are formed with groups of apolar, cationic or anionic aminoacids. Despite its low isoelectric point, HSA can also complexate neutral and negatively charged molecules, binding the broadest known variety of chemical structures that interact with proteins. Furthermore, on top of its globally rigid structure, HSA shows a remarkably dynamic conformation, being able to change its local structure in order to accommodate different molecules. Nonetheless, ligand binding is noticeably specific, i.e. a particular binding site can discriminate between similar structure or even between enantiomers of the same molecule. At least six binding areas has been characterized on HSA. The first two areas bind several common drugs, the third fatty acids, the fourth bilirubin and dyes, the fifth divalent cations and the sixth haemin. The drug-binding areas are remarkably specific; the first one binds salicylates, coumarines such as warfarin, and pyrazolidines such as phenylbutazone, with high affinity. The second one strongly interacts with arylpropionic acids such as ibuprofen, benzodiazepines such as diazepam, and indoles such as tryptophan. Moreover, the same drug can bind to either primary high-affinity sites or lower-affinity secondary sites, the second ones generally in a less specific fashion. In general, the same protein area can act as a primary site for some ligands and as a secondary site for other ones.

The overall structure of HSA is known, and also an X-ray three-dimensional model has been reported, although under extremely non-physiological conditions (the crystals were grown in the presence of polyethylen-glycol; He and Carter, 1992). Nonetheless, due to the complexity and dynamic variety of its interactions, HSA is still being studied. Also thanks to its commercial availability, HSA is considered a general model for the approach to the analysis of the relationship between protein structures and activity.

Several conformational isomers of HSA interconvert *in vivo* depending on temperature, pH and ionic strength (Honorè, 1990). Under physiological conditions, the average secondary structure is composed of over 60% of α-helix, about 16% of β-structures and the remaining portion in random coil. The fine distribution of substructures is indeed relevant to determine the subdomains arrangement, and thus binding activity (He and Carter, 1992).

Post-translational modifications and secondary structure

In order to study the influence of local modifications on the secondary structure, several post-translational modifications were induced *in vitro* (fig. 80).

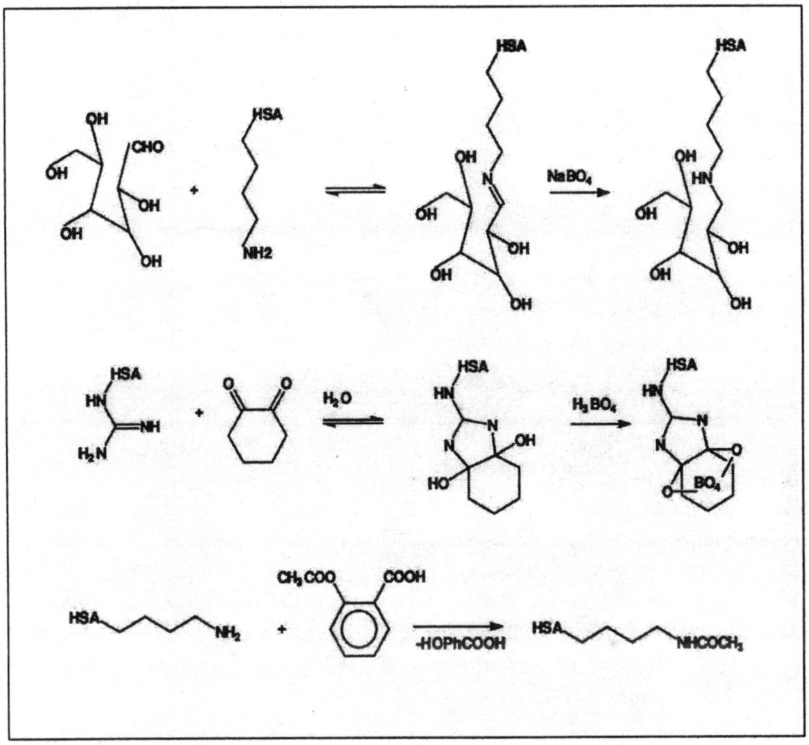

Figure 80. *HSA* in vitro *post-translational modifications: glycosylation, Amadori's dichetonization and aspirin-induced acetylation.*

The acetylation induced by aspirin is a well-known reaction (Walker, 1976), occurring *in vivo* after ingestion of a normal dosage of the most common drug. It is a site-specific reaction in that aspirin binds to the salicylate site (area I) and slowly transfers its acetyl group to an internal lysine of HSA. The dichetonization is a chemical reaction normally used to block protein arginines (e.g. Hugli, 1989), while the glycosylation is another non-catalyzed physiological reaction particularly relevant in diabetic individuals (Kragh-Hansen, 1990). Several samples of glycosylated HSA, different for incubation time, were prepared (Viegi, 1994).

Short wavelength CD spectra of HSA and all the modified forms were recorded after SDS-PAGE analysis and without further purification (fig. 81).

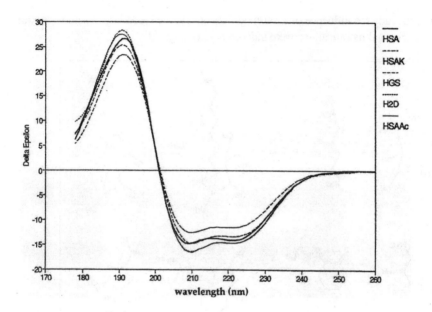

Figure 81. *CD spectra of native human serum albumin (HSA) and its modified forms: dichenonized (HSAK), commercial glycosylated sample (HGS), in vitro glycosylated (H2D, 8 weeks of incubation with glucose) and acetylated (HSAAc); c=1 mg/ml, cell pathlength 0.0095 cm.*

The deconvolution analysis was performed according to the method of Johnson (see Introduction and Chapter 3), and the results are reported in tab. 11.

	H	A	P	T	O	
HSA	0.66±0.01	-0.03±0.01	0.03±0.01	0.13±0.02	0.22±0.02	1.01
HSAGs	0.66±0.01	0±0.02	0.01±0.01	0.08±0.01	0.24±0.01	0.99
HSAAc	0.68±0.01	-0.05±0.01	-0.03±0.01	0.12±0.01	0.21±0.01	1.00
H2D	0.61±0.01	0.02±0.05	-0.01±0.01	0.14±0.01	0.21±0.02	0.98
HSAK	0.59±0.01	-0.03±0.01	0.02±0.01	0.14±0.01	0.25±0.01	0.99

Table 11. *Secondary structures of HSA and its modified forms (symbols as in fig. 81) calculated from data of fig. 81: H, α-helix; A, antiparallel β-sheet; P, parallel β-sheet; T, β-turn; O, other forms and random coil. The sum of the components (last column) is an index of the accuracy of the analysis.*

It is interesting to observe that, while the local and minimal modification induced by aspirin produces virtually no effects on the secondary structure, the extensive dichetonization, albeit reported to involve mainly a single residue of arginine (Peters, 1985) denaturates HSA to a much deeper extent, while the influence of glycosylation depends, as expected, on the extent of the modification. From these data, it is possible to foresee a drastic change in binding protein of dichetonized HSA with respect to the native and acetylated protein.

Binding properties of native and modified HSA: a spectroscopic study.

Stereospecificity is one of the most peculiar properties of ligand binding to proteins in general and to HSA in particular. A large number of works has been published by several authors (for examples, Fehske et al., 1981; Fitos et al., 1986; Maruyama et al., 1990), and a contribution to the spectroscopic study of this problem is here presented (see also Ascoli, 1993; Bertucci et al., 1995a and 1995b; Ascoli et al., 1995). In fact, HSA ligand binding is such a wide subject that even the analysis of its stereospecific interactions represents a large chapter of the study of the relationship between structure and activity. The spatial disposition of protein binding areas allows not only to discriminate between two enantiomers of a chiral molecule, but also to bind a prochiral drug in an optically active conformation. In the present appendix, only a single example will be presented, i.e. the spectrometric study of the interaction of HSA with prochiral diazepam (DZP), the most representative (and world-wide most prescribed) benzodiazepine.

When DZP forms a complex with HSA, its solvent environment undergoes a polarity change, as DZP passes from a water solution into a more apolar binding site on the protein. This effect causes a red shift of the DZP's UV absorption band at low energy, where HSA has virtually no absorption. Therefore, by subtracting from the complex spectrum those of the two components, a difference UV (ΔUV) spectrum is obtained, which measures the extent of the interaction. Since the red shift phenomenon is a general effect of ligand/protein interaction, every kind of binding is revealed by this method, i.e. both high-affinity and low-affinity complexations. It is possible to titrate the binding sites by adding increasing amounts of ligand to a solution of HSA, and recording the ΔUV spectra (fig. 82).

In contrast, CD spectrum only detects optically active molecules, thus free prochiral DZP has no dichroic bands in solution. When it binds to protein stereospecific sites, DZP assumes a chiral conformation and shows as a consequence a CD signal corresponding to its UV absorptions. The low-energy band (in a spectral region where HSA has virtually no dichroic absorptions) can be used to measure stereospecific interactions, and ΔCD titration curves can be built (fig. 83).

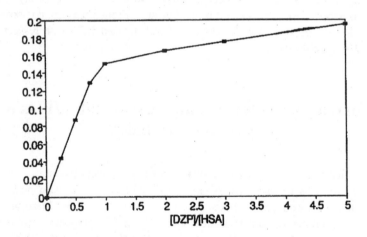

Figure 82. *Difference UV (A_{261}-A_{250}) titration of HSA/DZP complex. [HSA] = 20 μM, cell pathlength = 1 cm.*

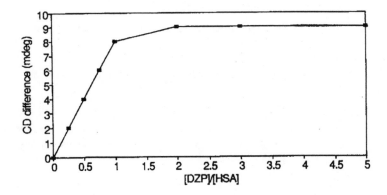

Figure 83. *ΔCD titration of the HSA/DZP complex. Same conditions as in fig. 82.*

It appears evident that, while a single high-affinity stereospecific binding site is present, several other low-affinity binding areas are also detected, which complexate DZP in an optically-inactive conformation. This interpretation is in perfect agreement with literature data.

A ΔCD method to analyze quantitatively the stereospecific interactions was also developed by means of dilution curves; according to the Lambert-Beer law, if a complex is diluted and the spectrometric cell pathlength is proportionally increased, the resulting decrease of the signal is uniquely due to the entropy effect of dilution, which depends of the affinity constant of the binding site (fig. 84).

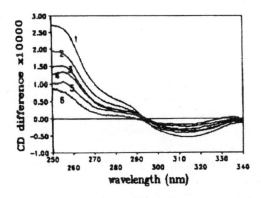

Figure 84. *Dilution curve of 1/1 HSA/DZP complex ([1], c = 1 mM, l = 0.01 cm; [2], c = 0.2 mM, l = 0.05 cm; [3], c = 50 μM, l = 0.2 cm; [4], c = 5 μM, l = 2 cm; [5], c = 2 μM, l = 5 cm; [6], c = 1 μM, l = 10 cm).*

For CD analysis, which only measures the stereospecific interactions, the following linear equation was developed for a stoichiometric interaction 1/1 (Ascoli et al., 1995):

$$p \; l^{1/2} \; CD^{-1/2} = \Delta\varepsilon^{-1} \; l^{-1/2} \; CD^{1/2} + K^{-1/2} \; \Delta\varepsilon^{-1/2}$$

being p the concentration of the protein, CD the measured signal, l the cell pathlength, K the affinity constant and $\Delta\varepsilon$ the molar ellipticity. By changing p and l, and measuring CD, it is possible to obtain K and $\Delta\varepsilon$ by linear fitting, i.e. to measure the affinity and the stereospecificity of the interaction (fig. 85).

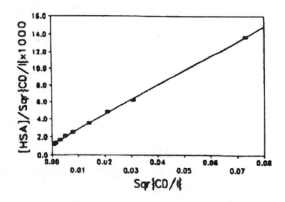

Figure 85. *Linear analysis of data from fig. 84 (x units: $cm^{-1/2}$; y units: $M\ cm^{1/2}$). Slope=0.172; intercept=1.11·10^{-3}. The values of $140\pm10\ 10^{3}\ M^{-1}$ and $5.81\pm0.35\ M^{-1}cm^{-1}$ are obtained for K and $\Delta\varepsilon$, respectively.*

The main advantages of this method are simplicity and velocity, in addition to the fact that the preliminary separation between macromolecules and free ligands, necessary for classical biochemical methods such as dialysis, is also avoided. Besides, no other methods have been developed so far to detect quantitatively stereospecific interactions.

This technique was also extended to study competitive interactions, i.e. the total or partial displacement of a ligand when other molecules are simultaneously bound (cobinding). In particular, we showed that two distinct cobinding phenomena can be analyzed, namely direct and indirect competition. Direct competition was performed with DZP and ibuprofen (fig. 86), an antiinflammatory drug that binds to the same area as DZP (HSA binding area II).

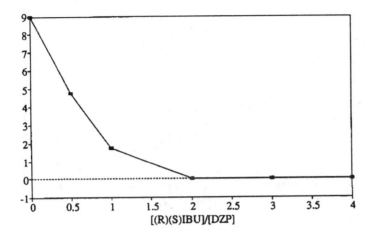

Figure 86. *Direct competition of the DZP/HSA complex(-CD difference at 315 nm, in mdeg): displacement by ibuprofen.*

Indirect competitive binding was studied with salicylate, that binds to area I on HSA (fig. 87).

Direct and indirect competition may be distinguished by CD because only direct displacers reduce the induced signal of the probe to zero (figures 86 and 87). This effect was also confirmed with a drug binding at area I, phenylbutazone (Ascoli et al., 1995).

A case of independent binding was also studied, using the chemoterapic drug fluorouracil (fig. 88), which has a secondary, low-affinity binding site in HSA area I.

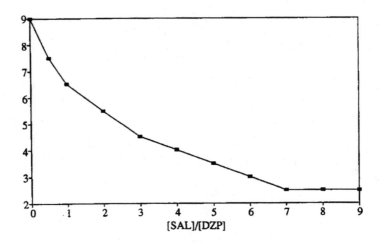

Figure 87. *Indirect competition of the DZP/HSA complex: displacement by salicylate.*

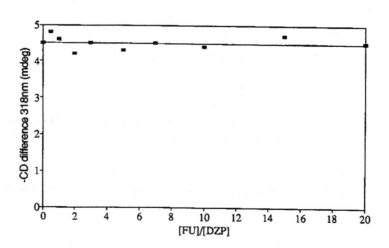

Figure 88. *Independent binding of DZP (area II) and fluorouracil (primary site in an area other than I and II, secondary site in area I) to HSA. The induced CD signal of prochiral DZP is not reduced.*

Competitions were quantitatively analyzed as well, and equations developed to calculate affinity constants for either direct or indirect displacers (Ascoli et al., 1995), on the basis of the affinity and spectroscopic data of the probe, (K and $\Delta\varepsilon$), which can be measured as in fig. 85. The data obtained for typical competitions at areas I and II are reported in tab. 12, but the detailed description of the equations is beyond the aim of this appendix and in general of a model study.

Ligand	K(I), 1/M	K(II), 1/M
PHE	84000	n.d
DZP	n.d.	141000
SAL	16000	13000
IBU	36000	140000

Table 12. *Affinity constant calculated for the stereospecific binding of diazepam, phenylbutazone, ibuprofen and salicylate at areas I and II; n.d., not determined.*

It should be emphasized that also CD analysis specifically detects interactions at stereoselective binding areas also in the case of competitions.While this represents a unique advantage, other methods such as UV spectroscopy or biochemical techniques must be employed for the study of all non-stereospecific interactions.

In order to obtain insights into the influence of post-translational modification on protein activity, binding studies were carried out in parallel on modified albumins. The dichetonised HSA showed a complete disruption of the stereospecific binding site, while preserving the non-specific interactions as detected by ΔUV. In contrast, the titration curves for acetylated and glycosylated HSA were similar to those obtained for native HSA (fig. 89).

As expected, an impaired binding activity is observed in samples whose secondary structure had drastically changed.

In contrast, marked differences between native HSA and the acetylated protein were observed with competition studies. The direct displacement of DZP by ibuprofen is shown in fig. 90.

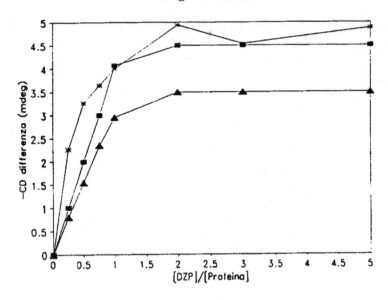

Figure 89. *Ellipticity at 315 nm (in -mdeg) of 1/1 complexes DZP/human serum albumin. HSA, squares; HSAAc, triangles; H2D, asterisks. The corresponding curve for HSAK was below 0.5 mdeg over the entire titration range.*

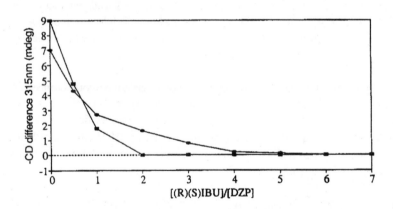

Figure 90. *Ibuprofen-caused reduction of DZP's induced ΔCD measured on native HSA (squares) and acetylated HSAAc (crosses).*

The indirect competition by salicylate is reported in fig. 91.

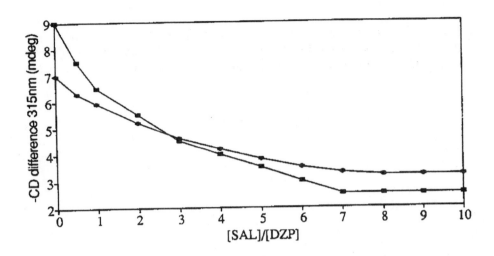

Figure 91. *Salicylate caused reduction of DZP's induced ΔCD measured on native HSA (squares) and acetylated HSAAc (crosses).*

The effect of acetylation is a minor displacement efficiency for both direct and indirect competitors. This phenomenon is directly caused by the local post-translational modification; the acetylation transforms an amminic group in an ammidic one, and the polarity of the environment is drastically reduced. In particular, the lys_{199} amino group (protonated and thereby positively charged at physiological pHs) turns into a neutral functional group. While diazepam is a neutral apolar molecule, both competitors ibuprofen and salicylate are negatively charged under normal conditions. It is thus to be expected that the displacement power is reduced after such particular acetylation. A similar result is obtained with the apolar drug phenylbutazone binding at area I (Bertucci et al., 1995a). This also confirms that lysine 199 contributes to the formation of both binding areas I and II, as postulated on the basis of X-ray structure (He and Carter, 1992).

Experiment are in progress to characterize the binding properties of the HSA glycosylated samples.

A parallel study of HSA and Ac-HSA binding of fluorouracil was also carried out with these techniques and with nuclear magnetic resonance, suitable to measure (on a single sample) the ^{19}F-band enlargement of the ligand and the changes of the protein's proton signals. A secondary interaction with HSA (but not Ac-HSA) binding area I was detected (Bertucci et al., 1995b) demonstrating that an integrate spectroscopical approach can furnish insights into the binding mechanism at the levels of both the ligand and the protein.

Several attempts were made to analyze the HSA post-translational modifications with IS-MS, but only the exact molecular weights of the native cloned protein and of bovine serum albumin (BSA) could be measured with satisfactory accuracy.

C.   Post-translational modifications of B-50 by MS.

B-50, also termed GAP-43, F1, and neuromodulin, has been extensively investigated in the past 15 years (for a review see e.g. Gispen at al., 1991). This protein, a nervous tissue specific proteic substrate of PKC, is involved in neuronal growth cone functioning (Van Hooff et al., 1988), transmembrane signal transduction mechanisms (De Graan et al., 1988a) and synaptic potentiation (De Graan et al., 1988b, Gianotti et al., 1992). B-50 was characterized as an acidic protein with isoelectric point 4.3-4.7, showing anomalous behavior in SDS-PAGE, since its apparent molecular weight variates with increasing concentration of acrylamide (apparent 43-57 kD, real 23-25 kD). B-50 is an elongated and extremely hydrophilic molecule of 226 to 247 aminoacids depending upon the species, particularly enriched in glycine, glutamate and aspartate residues and relatively rich in alanine and histidine. Even if no hydrophobic domains are present in the sequence, the B-50 protein is localized primarily on the cytoplasmatic side of plasma membrane of nerve terminals. B-50 is anchored to the synaptic membrane via a fatty acylation (most likely a palmitoylation) on the N-terminal cysteins in position 3 and 4. This is thus an interesting example of compartmentalization dependent on post-translational modifications.

On the B-50 sequence, different functional domains have been identified: a Calmodulin (CAM) binding domain becomes inactive when an inner domain (serine 41) is phosphorylated by PKC. At least two serines phosphorylatable by other kinases are also present, although no functional *in vivo* role has been found yet for these non-PKC mediated reactions. The Calmodulin binding function of dephospho-B-50 is a crucial activity since Calmodulin is a local storage of Ca^{2+}, and also an effector of another protein kinase pathway (CAM-kinase). Furthermore, it has been unequivocally proven that the modulation of B-50 phosphorylation plays a central role in the regulation of growth cones, which lead axonal extremities when they reach out to dendrites or dendritic spines in order to form new synapses. This effect is important for both neural development and memory, and is one of the most important problems of synaptic plasticity. It also confirms that development (biological memory) and synaptic plasticity (neural memory) share common physiological and biochemical pathways.

Besides phosphorylation, other post-translational modifications have been reported for B-50, such as ADP-rybosylation (Palkiewicz et al., 1994), and although the functional relevance of some of these modifications still awaits further clarification, the interest in a complete analysis of B-50 post-translational modifications *in vivo* and in *vitro* is evident. A general method is necessary to determine quickly the degree of each modification in order to establish a correlation with physiological conditions.

B-50 is member of a group of neural PKC substrates which share common structural and biochemical properties, e.g. they are soluble in perchloric acid. One of the most interesting elements of this group aside from B-50 is neurogranin (also known as p17,

BICKS or RC3; Baudier et al, 1991), which also binds Calmodulin and shows sequence homology with B-50 in a short domain of 15 aminoacids, including the Calmodulin/PKC binding domain. Immunocytochemical studies revealed a predominant staining for B-50 in presynaptic nerve terminals (Gispen et al., 1985) and for neurogranin in the perikarya and the dendrites of neurons (Represa et al., 1990). Although its physiological role is still to be unraveled, neurogranin is present, like B-50, in the cortex and hyppocampus, and its causal involvement with Long Term Potentiation (a widely accepted model of memory) has recently been proposed (Fedorov et al., 1995).

Purification and MS analysis of protein extracts

Standard methods for B-50 purification are based on an extraction from synaptosomal plasma membranes SPMs) isolated by several gradient ultracentrifugation steps (De Graan et al. 1993); briefly, SPMs are washed to release most of the loosely bound proteins, and the remaining pellet is extracted with 2-merchaptoethanol, a reducing agent which breaks the cystynic bond with the palmitoyl group anchoring B-50 to membranes. The supernatant is then subjected to a further purification step in a Calmodulin-sepharose column and B-50 is eluted in high calcium buffer. This method therefore yields exclusively dephospho- and depalmitoyl-B-50. Methods currently available to measure the degree of phosphorylation of a protein are largely based upon radioenzymatic procedures involving back-titration to label the number of its *in vitro* phosphorylatable sites by means of ^{32}P-phosphate incorporated into the residual dephospho-form. An alternative to this post-hoc phosphorylation assay is based on ^{32}P-ortho-phosphate labeling of excised tissues and subsequent immunoprecipitation of the protein of interest. None of these methods gives satisfactory results in case of purified B-50 due to its peculiar isolation procedure, and B-50 was thus one of the first neuronal proteins on which a mass spectrometry approach was attempted (Di Luca et al., 1992). In particular HPLC-IS/MS can detect phospho- and dephospho-forms of a native protein, as well as other post-translational modifications, in the same sample, furnishing a direct analytical tool, as discussed in the Introduction.

Acidic extracts of rat brain were analyzed by HPLC-MS. By eluting the extracts onto on-line C18-reverse phase-HPLC (see e.g. Biemann, 1992) neurogranin was detected by IS-MS (fig. 92).

Another protein, identified as ubiquitin, was also revealed by MS (fig. 93).

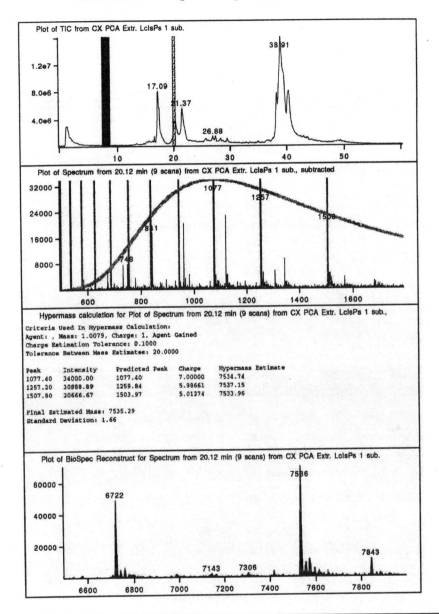

Figure 92. *Single rat hyppocampus extract C18 RP-HPLC/IS-MS: total ion current chromatogram (first panel from the top); mass spectrum of the area shaded in gray after subtraction of the area shaded in black (second panel); deconvolution of mass and charge, which identifies the protein as phosphorylated neurogranin (third panel); reconstructed MW spectrum (bottom).*

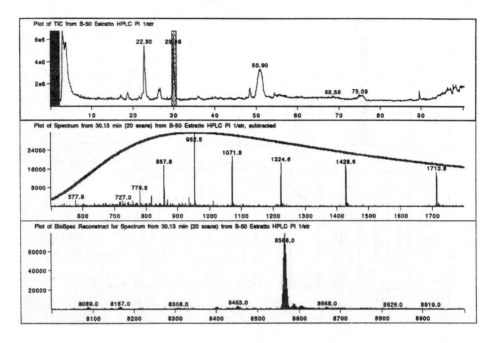

Figure 93. *Identification of Ubiquitin by C18 RP-HPLC/IS-MS from single rat hyppocampus extract. Legend as in fig. 91.*

The protein recovery from the C18 column was extremely poor. However, it is possible to detect the phospho- and dephospho- forms of both B-50 and neurogranin as well as ubiquitin in the same injection by means of an on-line perfusion chromatography (Di Luca, et al., 1996).

The list of the peaks detected by on-line HPLC/IS-MS is reported in tab. 13.

One of the aims of the present research project was the study of phosphorylation of neural proteins by MS techniques. Since preliminary results indicated that both phospho- and dephospho-B-50 are detectable from acidic extracts, an experiment was set-up to phosphorylate *in vitro* proteins extracted by PKC.

PKC was purified from rat brain according to standard procedures (Kikkawa et al., 1986), activated by phosphatidylserine and calcium, and incubated with ATP and freshly perchloric acid extracted proteins from rat cortex after neutralization and desalting.

Ret. time	MW	Abundance	Assignment	Predicted
12,10	**4962**	**80000**	**n.d.**	**n.d.**
14,20	7450	28000	neurogranin	7495
14,20	**7535**	**40000**	**P-neurogr.**	**7575**
14,80	**6719**	**55000**	**n.d.**	**n.d.**
16,10	**8565**	**25000**	**ubiquitin**	**8565**
16,10	8617	25000	n.d.	n.d.
17,60	**12975**	**47000**	**n.d.**	**n.d.**
18,30	14687	70000	n.d.	n.d.
27,00	23602	12000	B-50	23602
27,00	23682	5000	P-B-50	23682
33,20	22030	7000	n.d.	n.d.
36,00	33619	5000	n.d.	n.d.

Table 13. *MW reported for rat brain acid extract by HPLC/MS. Boldface-typed peaks were also observed by RP-HPLC/IS-MS. Predicted mass is based on cDNA sequences. Abundance is based on total ion current. n.d.= not determined.*

An aliquot of the samples was also incubated with ^{32}P-ATP and the SDS-PAGE analysis of the resulting mixture clearly demonstrated the occurrence of protein phosphorylation *in vitro* (fig. 94). In particular, three bands can be assigned to PKC auto-phosphorylation, B-50 and neurogranin.

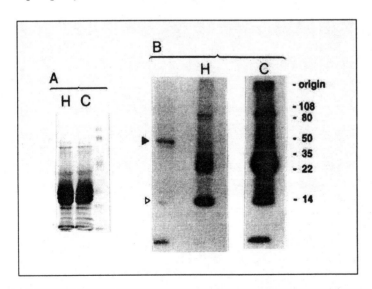

Figure 94. *(A) Acidic extract western blot. (B) Back-phosphorylation: ^{32}P-bands after incubation with PKC, Ca^{2+}/phosphatidylserine, ^{32}P-ATP. H=hippocampus, C=cortex. Triangles: purified B-50 (control) and neurogranin position.*

In order to test the relevance of other post-translational modification such as palmitoylation and ADP-rybosylation, some *in vitro* experiments were also performed on acidic extracts as well as on cloned and on purified B-50. ADP-rybosylation was carried out both non-enzimatically and with an incubation mixture containing SPMs and/or toxins known to induce this modification in other proteins (M.P. Abbracchio, Inst. Pharmacol. Univ. Milan, Italy, personal communication). Palmitoylation was also tried both in catalyzed and non-catalyzed conditions, and analysis by western blot and radioactive SDS-PAGE indicated that both post-translational modifications are likely to occur *in vitro*. Experiments are in progress to analyze the samples with on-line HPLC-IS/MS and to investigate the physiological meaning of these reactions.

REFERENCES

Acosta-Urquidi, J.; Alkon, D.L.; Neary, J.T.: Ca^{2+}-dependent protein kinase injection in a photoreceptor mimics biophysical effects of associative learning. *Science* **224**, 1254-1257 (1984).

Akers, R.F.; Routtemberg, A.: Protein Kinase C phosphorylates a 47 Mr protein (F1) directly related to synaptic plasticity. *Brain Res.* **334**, 147-151 (1985).

Akke, M.; Forsèn, S.; Chazin, W.J.: Solution structure of $(Cd^{2+})_1$-calbindin D_{9k} reveals details of the stepwise structural changes along the apo $\rightarrow (Ca^{2+})_1^{II} \rightarrow (Ca^{2+})_2^{I,II}$ binding pathway. *J. Mol. Biol.* **252**, 102-121 (1995).

Alkon, D.L.: Associative training of Hermissenda. *J. Gen. Physiol.* **64**, 70-84 (1974).

Alkon, D.L.: Voltage dependent calcium and potassium ion conductance: a contingency mechanism for an associative learning model. *Science* **205**, 810-813 (1979).

Alkon, D.L.; Lederhendler, I.; Shoukimas, J.L.: Primary changes of membrane currents during retention of associative learning. *Science* **215**, 693-695 (1982).

Alkon, D.L.; Acosta-Urquidi, J.; Olds, J.L.; Kuzma, G.; Neary J.: Protein kinase injection reduces voltage-dependent potassium currents. *Science* **219**, 303-305 (1983)

Alkon, D.L.: Calcium-inactivated potassium currents: a biophysical memory trace. *Science* **276**, 1037-1045 (1984).

Alkon, D.L.; Sakakabara, M.; Forman, R.; Harrigan, J.; Lederhendler, I.; Farley, J.: Reduction of twp voltage-dependent K+ currents mediates retention of a learned association. *Behav. Neural Biol.* **44**, 278-300 (1985).

Alkon, D.L; Rasmussen, H.: A spatial and temporal model of cell activation. *Science* **239**, 998-1005 (1988).

Alkon, D.L.; Naito, S.; Kubota, M.; Chen, C.; Bank, B.; Swallwood, J.; Gallant, G.; Rasmussen, H.: Regulation of Hermissenda K^+ channels by cytoplasmatic and membrane-associated C-kinase. *J. Neurochem.* **51**, 903-917 (1988).

Alkon, D.L.: Memory storage and neural systems. *Scientific American*, July 42-50 (1989).

Alkon, D.L.; Nelson, T.J.N.: Specificity of molecular changes in neurons involved in memory storage. *FASEB J.* **4**, 1567-1576 (1990).

Alkon, D.L.; Ikeno, H., Dworkin, J.; McPhie, D.L.; Olds, J.L., Lederhendler, I.; Matzel, J.; Shreus, B.G.; Kuziran, A.; Collin, C.; Yamoah, E.: Contraction of neuronal branching volume: an anatomical correlate of pavlovian conditioning. *Proc. Natl. Acad. Sci. USA* **87**, 1611-1614 (1990).

Alkon, D.L.: *Memory's Voice*. Harper&Collin ed. NY (1993).

Alkon, D.L.; Blackwell, K.T.; Barbour, G.S.; Werness, S.A.; Vogl, T.P.: Biological plausibility of synaptic associative memory models. *Neural Networks* **7**, 1005-1017 (1994).

Alkon, D.L.: Molecular mechanisms of associative memory and their clinical implications. *Behav. Brain Res.* **66**, 151-160 (1995).

Amburgey, J.C.; Abildgaard, F.; Starich, M.R.; Shah, S.; Hilt, D.C.; Weber, D.J.: ^{1}H, ^{13}C and ^{15}N NMR assignments and solution secondary structure of rat Apo-S100β. *J. Biomol. NMR* **6**, 171-179 (1995).

Ames, J.B.; Tanaka, T.; Stryer, L.; Ikura, M.: Seconday structure of myristoylated recoverin determined by three-dimensional heteronuclear NMR: implications for the calcium-myristoyl switch. *Biochem.* **33**, 10743-10753 (1994).

Ames, J.B.; Porumb, T; Tanaka, T.; Ikura, M.; Stryer, L.: Amino-terminal myristoylation induces cooperative calcium binding to recoverin. *J. Biol. Chem.* **270**, 4526-4533 (1995a).

Ames, J.B.; Tanaka, T.; Ikura, M.; Stryer, L.: Nuclear magnetic resonance evidence for Ca^{2+}-induced extrusion of the myristoyl group of recoverin. *J. Biol. Chem.* **270**, 30909-30913 (1995).

Anderson, J.A.; Sutton, J.P.: A network of networks: computation and neurobiology. *Proc. Int. Neu. Net. Soc. Annu. Meeting.* **1**, 561-568 (1995).

Ascoli, G.: Studi sulle interazioni fra siero albumina umana e piccole molecole organiche. M.S. thesis, University of Pisa (1993).

Ascoli, G.; Bertucci, C.; Salvadori, P.: Stereospecific and competitive bining of drugs to human serum albumin: a difference circular dichroism approach. *J. Pharm. Sci.* **84**, 737-741 (1995)

Ascoli, G.; Goldin, R.: Coordinate systems for dendritic spines: a somatocentric approach. Submitted to *Complexity* (1996).

Ashendel, C.L.: The phorbol ester receptor: a phospholipid-regulated protein kinase. *Biochim. Biophys. Acta* **822**, 219-242 (1985).

Aurell, L.; Friberger, P.; Karlsson, G.; Claeson, G.: A new sensitive and highly specific chromogenic peptide substrate for factor Xa. *Thrombosis Res.* **11**, 595-609 (1977).

Baitinger, C.; Alderton, J.; Poenie, M.; Schulman, H.; Steinhardt, R.A.: Multifunctional Ca^{2+}/calmodulin-dependent protein kinase is necessary for nuclear envelope breakdown. *J. Cell Biol.*
111,1763-1773 (1990).

Baudier, J.; Deloulme, J.C.; Van Dorsselear, A.; Black, D.; Matthes, H.W.D.: Purification and characterization of a brain specific protein kinase C substrate, neurogranin (p17). Identification of a consensus amino acid sequence between neurogranin and neuromodulin (GAP-43) that corresponds to the protein kinase C phosphorylation site and the calmodulin-binding domain. *J. Biol. Chem.* **266**, 229-237 (1991).

Berridge, M.J.; Irvine, R.F.: Inositol trisphosphate, a novel second messenger in cellular signal transduction. *Nature* **312**, 315-321 (1984).

Bertucci, C.; Domenici, E.; Salvadori, P.: Stereochemical features of 1,4-benzodiazepin-2-ones bound to human serum albumin: difference CD and UV studies. *Chirality* **2**, 167-174 (1990).

Bertucci, C.; Viegi, A.; Ascoli, G.; Salvadori, P: Protein binding investigation by difference circular dichroism: native and acetylated human serum albumins. *Chirality* **7**, 57-61 (1995a).

Bertucci, C.; Ascoli, G.; Uccello-Barretta, G.; Di Bari, L.; Salvadori, P.: The binding of 5-fluorouracil to native and modified human serum albumin: UV, CD an ^{1}H and ^{19}F NMR investigation. *J. Pharm. Biomed. Anal.* **13**, 1087-1093 (1995b).

Biemann, K.: Mass spectrometry of peptides and proteins. *Annu. Rev. Biochem.* **61**, 977-1010.

Bliss, T.V.; Collingridge, G.L.: A synaptic model of memory: long-term potentiation in the hyppocampus. *Nature* **361**, 31-39 (1993).

Bloom, T.L.; Szuts, E.Z.; Eckberg, W.R.: Inositol triphosphate, inositol phospholipid metabolism, and germinal vescicle breakdown in surf clam oocytes. *Dev. Biol.* **129**, 532-540 (1995).

Bonnell, B.S.; Keller, S.H.; Vacquier, V.D.; Chandler, D.E.: The sea urchin egg jelly coat consists of globular glycoproteins bound to a fibrous fucan superstructure. *Dev. Biol.* **162**, 313-324 (1994).

Bramanti, E.; Benedetti, E.: Determination of the secondary structure of isomeric forms of human serum albumin by a particular frequency deconvolution procedure applied to Fourier transform IR analysis. *Biopol.* **38**, 639-654 (1996).

Cameron, L.A.; Poccia, D.L.: In vitro development of the sea urchin male pronucleus. *Dev. Biol.* **162**, 568-578 (1994).

Chait, B.T.; Kent, S.B.H.: Weighing naked proteins: practical, high-accuracy mass measurement of peptides and proteins. *Science* **257**, 1885-1894 (1992).

Chazin, W.J.: Releasing the calcium trigger. *Nature Struct. Biol.* **2**(9), 707-710 (1995).

Chen, C.S.; Poenie, M.: New fluorescent probes for protein kinase C. Synthesis, characterization, and application. *J. Biol. Chem.* **268**, 15812-15822 (1993).

Churchland, P.S.; Sejnowsky, T.: *The computational brain.* MIT Press (1992).

Ciapa, B.; Allemand, D.; Payan, P.: Effect of the phorbol ester 12-O-tetradecanoylphorbol-13-acetate (TTPA) upon membrane ionic exchanges in sea urchin eggs. *Exp. Cell Res.* **185**, 407-418 (1989).

Cimler, B.M.; Giebelhaus, D.H.; Wakim, B.T.; Storm, D.R.; Moon, R.T.: Characterization of murine cDNAs encoding P57, a neural-specific calmodulin binding protein. *J. Biol. Chem.* **262**, 12158-12163 (1987).

Colley, P.A.; Routtemberg, A: Long-term potentiation as synaptic dialogue. *Brain Res. Rev.* **18**, 115-122 (1993).

Collin, C.; Miyaguchi, K.; Segal, M: Dendritic spines in cultured hippocampal slices: correlating structure and function. Submitted to *J. Neurosci.* (1996).

Connors, J.H.; Olds, J.L.; Lester, D.S.; McPhie, D.L.; Senft, S.L.; Johnston, J.A.; Alkon, D.L.: Heterogeneous distribution of fluorescent phorbol ester signal in living sea urchin embryos. *Biol. Bull.* **183**, 407-418 (1992).

Costantino, H.R.; Griebenow, K; Mishra, P.; Langer, R.; Klibanov, A.M.: Fourier-transform infrared spectroscopic investigation of protein stability in the lyophilized form. *Biochim. Biophys. Acta* **1253**, 69-74 (1995).

Crow, T.J.; Alkon, D.L.: Retention of an associative behavioral change in Hermissenda. *Science* **201**, 1239-1241 (1978).

Crow, T.J.; Alkon, D.L.: Associative behavioral modification in Hermissenda: cellular correlates. *Science* **209**, 412-414 (1980).

Dayhoff, J.: *Neural Network Architectures*. 23-31, Van Nostrand Reinhold NU (1990).

Dayhoff, J.; Hameroff, S.; Lahoz-Beltra, R.; Swenberg, C.E.: Cytoskeletal involvement in neural learning: a review. *Eur. Biophys. J.* **23**, 79-93 (1994).

De Graan, P.N.E.; Dekker, L.V.; De Wit, M.; Schrama, L.H.; Gispen, W.H.: Modulation of B50 phosphorylation and polyphosphoinositide metabolism in synaptic plasma membrane by Protein Kinase C, phorbol diesters and ACTH. *J. Receptor Res.* **8**, 345-361 (1988a).

De Graan, P.N.E.; Heemskerk, F.M.J.; Dekker, L.V.; Melchers, B.P.C.; Gianotti, C.; Schrama, L.H.: Phorbol esters induce long- and short-term enhancement of B50 (GAP-43) phosphorylation in rat hyppocampal slices. *Neurosci. Res. Comm.* **3**, 175-182 (1988b).

De Graan, P.N.E.; Moritz, A.; De Wit, M.; Gispen, W.H.: Purification of B50 by 2-mercaptoethanol extraction from rat brain synaptosomal plasma membranes. *Neurochem. Res.* **18**, 875-881 (1993).

De Matteis, M.A.; Santini, G.; Kahn, R.A.; Di Tullio, G.; Luini, A.: Receptor and protein kinase C-mediated regulation of ARF binding to the Golgi complex. *Nature* **364**, 818-821 (1993).

Dell'Angelica, E.C.; Schleicher, C.H.; Santomè, J.A.: Primary structure and binding properties of calgranulin C, a novel S100-like calcium-binding protein from pig granulocytes. *J. Biol. Chem.* **269**, 28929-28936 (1994).

Detering, N.K.; Decker, G.L.; Schmell, E.D.; Lennarz, W.J.: Isolation and characterization of plasma membrane-associated cortical granules from sea urchin eggs. *J. Cell. Biol.* **75**, 899-914 (1977).

Di Luca, M.: An animal model of impaired cognitive function. PhD thesis, University of Milan (1992), and references therein.

Di Luca, M.; De Graan, P.N.E.; De Angelis, L.; Gispen ,W.H.; Cattabeni, F.: Measurement of relative amountsof phospho- and dephospho-B50 (GAP43) peptides by fast atom bombardment mass spectrometry. *FEBS Lett*. **301**, 150-154 (1992).

Di Luca, M.; Pastorino, L.; Raverdino, V.; De Graan, P.N.E.; Caputi, A.; Gispen, W.H.; Cattabeni, F.: Determination of the endogenous phosphorylation state of B-50/Gap-43 and neurogranin in different brain regions by electrospray mass spectrometry. *FEBS Lett.*, in press (1996).

Domenici, E.: Chiral recognition at 1,4-benzodiazepin-2-ones receptors. Ph.D. thesis, Scuola Normale Superiore of Pisa (1990).

Donaldson, C.; Barber, K.R.; Kay, C.M.; Shaw, G.S.: Human S100b protein: formation of a tetramer from synthetic calcium-binding site peptides. *Prot. Sci.* **4**, 765-772 (1995).

Dudai, Y.: *Neurobiology of learning and memory*. Oxford Press (1989).

Durussel, I.; Rilliet, Y.L.; Petrova, T.; Takagi, T.; Cox, J.A.: Cation binding and conformation of tryptic fragments of Nereis Sarcoplasmic calcium-binding protein: calcium-induced homo- and heterodimerization. *Biochem.* **32**, 2394-2400 (1993).

Eckberg, W.R.; Szuts, E.Z.; Carroll, A.G.: Protein kinase C activity, protein phosphorylation and germinal vesicle breakdown in Spisula Oocytes. *Dev. Biol.* **124**, 57-64 (1987).

Eckberg, W.R; Palazzo, R.E.: Regulation of M-phase progression in Chaetoptorus oocytes by protein kinase C. *Dev. Biol.* **149**, 395-405 (1992).

Eckberg, W.R.; Szuts, E.Z.: Diacylglycerol content of Chaetopterus ooyctes during maturation and fertilization. *Dev. Biol.* **159**, 732-735 (1993).

Eckberg, W.R.; Carroll, A.G.: Evidence for involvement of protein kinase C in germinal vesicle breakdown in Chaetopterus. *Develop. Growth and Different.* **29**, 489-496 (1995).

Epand, R.M.; Lester, D.S.: The role of membrane biophysical properties in the regulation of protein kinase C activity. *Trends Pharmacol. Sci.* **11**, 317-320 (1990).

Etcheberrigaray, R.; Ito, E.; Kim, S.; Alkon, D.L.: Soluble β-amyloid induction of Alzheimer's phenotype for human fibroblast K⁺ channels. *Science* **264**, 276-279 (1994).

Farley, J.; Richards, W.; Ling, L.J.; Liman, E.; Alkon, D.L.: Membrane changes in a single photoreceptor cause associative learning in Hermissenda. *Science* **221**, 1201-1203 (1983).

Farley, J.; Auerbach, S.: Protein kinase C activation induces conductance changes in Hermissenda photoreceptor like those seen in associative learning. *Nature* **319**, 220-223 (1986).

Federov, N.B.; Pasinelli, P.; Oestreicher, A.B.; De Graan, P.N.E.; Reymann, K.G.: Antibodies to post-synaptic PKC substrate neurogranin prevent Long Term Potentiation in hippocampal Ca1 neurons. *Eur. J. Neurosci.* **7**, 819-822 (1995).

Fehske, K.J.; Mueller, W.E.; Wollert, U.: The location of drug binding sites in human serum albumin. *Biochem. Pharrmacol.* **30**, 687-692 (1981).

Findlay, W.A.; Sykes, B.D.: ¹H-NMR resonance assignments, secondary structure, and global fold of the TR₁C fragment of turkey skeletal troponin C in the calcium-free state. *Biochem.* **32**, 3461-3467 (1993).

Fitos, I.; Tegyey, Zs.; Simonyi, M.; Kajtar, M.: Stereoselective allosteric interaction in the binding of lorazepam methylether and warfarin to human serum albumin. In "Bio-organic Heterocycles 1986 - Synthesis, mechanisms and bioactivity", Proc. 4ᵗʰ FECHEM Conf. Van der Plas, H.C.; Simonyi, M.; Alderweireldt, F.C.; Lepoivre, J.A. eds Elsevier (1986).

Fu, F.-N.; DeOliveira, D.B.; Trumble, W.R.; Sarkar, H.K.; Singh, B.R.: Secondary structure estimation of proteins using the amide III region of Fourier transform infrared spectroscopy: application to analyze calcium-binding-induced structural changes in calsequestrin. *Appl. Spectr.* **48**, 1432-1441 (1994).

Gardner, D.: *The Neurobiology of Neural Networks.* 21-71, MIT Press (1993).

Geourjon, C.; Deleage, G.: SOPM: a self optimised prediction method for protein secondary structure. *Prot. Eng.* **7**, 157-164 (1994).

Geourjon, C.; Deleage, G.: Significant improvements in protein secondary structure prediction from multiple alignments. *Comput. Applic. Biosci.* **11**, 681-684 (1995).

Gianotti, C.; Nunzi, M.G.; Gispen, W.H.; Corradetti, R.: Phosphorylation of the presynaptic B-50 (GAP-43) is increased during electrically induced long-term potentiation. *Neuron* **8**, 843-848 (1992).

Gillot, I.; Ciapa, B.;, Payan, P.; De Renzis, G.; Nicaise, G.; Sardet, C.: Quantitative X-ray microanalysis of calcium in sea urchin eggs after quick-freezing and freeze-substitution. Validity of the method. *Histochem.* **92**, 523-529 (1989).

Gispen, W.H.; Leunissen, J.L.M.; Oestreicher, A.B.;Verkleij, A.J.; Zwirs, H.: Presynaptic localization of B50 phosphoprotein: the ACTH-sensitive protein kinase and its substrate in rat brain membranes. *Brain Res.* **328**, 381-385 (1985).

Gispen, W.H.; Nielander, H.B.; De Graan, P.N.E.; Oestreicher, A.B.; Schrama, L.H.; Shotman, P.: Role of the growth associated protein B-50/GAP-43 in neuronal plasticity. *Mol. Neurobiol.* **5**, 61-85 (1991).

Gorenstein, C.; Bundman, M.C.; Lew, P.J.; Olds, J.L., Ribak, C.E.: Dendritic transport. i. colchicine stimulates the transport of lysosomal enzymes from cell bodies to dendrites. *J. Neurosci.* **5**, 2009-2017 (1985).

Hadden, J.M.; Chapman, D.; Lee, D.C.: A comparison of infrared spectra of proteins in solution and crystalline forms. *Biochim. Biophys. Acta* **1248**, 115-122 (1995).

Hagstrom, B.E.; Lonning, S.: The sea urchin egg as a testing object in toxicology (Abstract). *Acta Pharm. et Tox.* **32**, 7-49 (1973).

Harris, K.M., Kater, S.B.: Dendritic spines: cellular specializations imparting both stability and flexibility to synaptic functions. *Annu. Rev. Neurosci.* **17**, 341-371 (1994).

Harris, K.M.: How multiple synapse boutons could preserve input specificity during an interneuronal spread of LTP. *Trends Neurosci.* **18,** 165-169 (1995).

Hazelbauer, G.L.; Harayama, S.: Sensory transduction in bacterial chemotaxis. *Int. Rev. Cytol.* **81**, 33-70 (1983)

He, X.M.; Carter, D.C.: Atomic structure and chemistry of human serum albumin. *Nature* **358**, 209-215 (1992).

Hennessey, J.P.; Johnson, W.C.Jr.: Information content in the circular dichroism of proteins. *Biochem.* **20**, 1085-1094 (1981).

Hennessey, J.P.Jr.; Manavalan, P.; Johnson, W.C.Jr.; Malencik, D.A.; Anderson, S.R.; Schimerlik, M.I.; Shalitin, Y.: Conformational transition of calmodulin as studied by vacuum-UV CD. *Biopol.* **26**, 561-571 (1987).

Hinegardner, R.: Care and handling of sea urchin eggs, embryos, and adults (principally north american species). *The Sea Urchin Embryo* (Czihak, G. ed.), 10-22. Springer-Verlag Berlin (1975).

Honorè, B.: Conformational changes in human serum albumin induced by ligand binding. *Pharmacol. Toxicol.* **66**, 7-26 (1990).

Hosokawa, T., Rusakov, D.A.; Bliss, T.V.P., Fine. A: Repeated confocal imaging of individual dendritic spines in the living hippocampal slice - Evidence for changes in length and orientation associated with chemically-induced LTP. *J. Neurosci.* **15**, 5560-5573 (1995).

Huang, E.C.; Pramanik, B.N.; Tsarbopoulos, A.; Reichter, P.; Ganguly, A.K; Trotta, P.P.; Nagabhushan, P.L.; Covey, T.R.: Application of electrospray mass spectrometry in probing protein-protein and protein-ligand noncovalent interactions. *J. Am. Soc. Mass Spectr.* **4**, 624-630 (1993).

Huber, L.A.; de Hoop, M.J.; Dupree, P.; Zerial, M.; Simons, K.; Dotti, C.: Protein transport to the dendritic plasma membrane of cultured neurons is regulated by rab8p. *J. Cell. Biol.* **123**, 47-55 (1993).

Hugli, T.E.: *Techniques in protein chemistry.* Academic Press Boston (1989).

Husten, E.J.; Eipper, B.A.: Purification and characterization of PAM-1, an integral membrane protein involved in peptide processing. *Arch. Biochem. Biophys.* **312**(2), 487-492 (1994).

Ikura, M.: Calcium binding and conformational response in EF-hand proteins. *Trends Biochem. Sci.* **21**, 6-9 (1996).

Johansson, C.; Ullner, M.; Drakenberg, T.: The solution structures of mutant calbindin D_{9k}'s, as determined by NMR, show that the calcium-binding site can adopt different folds. *Biochem.* **32**, 8429-8438 (1993).

Johnson, W.C.Jr.: Protein secondary structure and circular dichroism: a practical guide. *Proteins* **7**, 205-214 (1990).

Johnson, W.C.Jr.: Analysis of circular dichroism spectra. *Methods in Enzymology* **210**, 426-447 (1992).

Johnston, J.A.; Sloboda, R.D.: A 62-kD protein required for mitotic progression is associated with the mitotic apparatus during M-phase and with the nucleus during interphase. *J. Cell. Biol.* **119**, 843-854 (1992).

Kandel, E.R.: *Cellular basis of behavior.* Freeman Ed. (1976).

Kandel, E.R.; Shwartz, J.H.; Jesser, M.: *Principles of Neuroscience.* 5th ed. Oxford (1993).

Kazemerak, L.: PKC and ion channels. In *Protein Kinase C current concepts and future perspective.* D.S Lester and R.M. Epand Eds. Ellis Horwood (1992).

Kikkawa, U.; Go, M.; Kuomoto, J.; Nishizuka, Y.: Rapid purification of PKC by HPLC. *Biochem. Biophys. Res. Commun.* **135**, 636-643 (1986).

Kim, C.S.; Han, Y.-F.; Etcheberrigaray, R.; Nelson, T.J.; Olds, J.L.; Yoshioka, T.; Alkon, D.L.: Alzheimer and b-amyloid-treated fibroblasts demonstrate a decrease in a memory-associated GTP-binding protein, cp20. *Proc. Natl. Acad. Sci. USA* **92**, 3060-3064 (1995).

Kragh-Hansen, U.: Structure and ligand binding properties of human serum albumin. *Dan. Med. Bull.* **37**, 57-84 (1990).

Lester, D.S.: In vitro linoleic acid activation of protein kinase C. *Biochim. Biophys. Acta* **1054**, 297-303 (1999).

Lester, D.S.; Alkon, D.L.: Activation of protein kinase c phosphorylation pathways: a role for storage of associative memory. *Progr. Brain Res.* **89**, 235-248 (1991).

Lester, D.S.: Membrane-associated protein kinase C. *Protein Kinase C: Current Concepts and Future*
Perpsectives (Lester, D.S., Epand, R.M. eds.) 80-101. Ellis Horwood Chichester (1992)

Li, J.M.; Parsons, R.A.; Marzluff, W.F.: Transcription of the sea urchin U6 gene in vitro requires a TATA-like box, a proximal sequence element, and sea urchin USF, which binds an essential E box. *Mol. Cell. Biol.* **14**, 2191-2200 (1994).

Linden, D.J.; Routtemberg, A.: The role of protein kinase C in long term potentiation: a testable model. *Brain Res. Rev.* **14**, 279-296 (1989).

Linse, S.; Brodin, P.; Drakenberg, T.; Thullin, E.; Sellers, P.; Elmaden, K.; Grundstroem, T.; Forsen, S.: Structure-function relationships in EF-hand Ca^{2+}-binding proteins. Protein engineering and biophysical studies of calbindin D_{9k}. *Biochem.* **26**, 6723-6735 (1987).

Lovinger, D.M.; Routtemberg, A.: Synapse-specific protein kinase C activation enhances maintanence of long-term potentiation in rat hyppocampus. *J. Physiol.* **400**, 321-333 (1988).

Lu, G.; Sehnke, P.C.; Feri, R.J.: Phosphorylation and calcium binding properties of an arabidopsis GF14 protein homolog. *Plant Cell* **6**, 501-510 (1994),

Lundberg, S.; Bjoerk, J.; Loefvenberg, L; Backman, L.: Cloning, expression and characterization of two putative calcium-binding sites in human non-erythroid α-spectrin. *Eur. J. Biochem.* **230**, 658-665 (1995).

Malenka, R.C.; Ayoub, G.S.; Nicoll. R.A: Phorbol esters enhance transmitter release in rat hyppocampal slices. *Brain Res.* **403**, 198-203 (1987).

Manning, M.C.: Unerlying assumptions in the estimation of secondary structure content in proteins by circular dichroism spectroscopy - a critical review. *J. Pharm. Biomed. Anal.* **7**, 1103-1119 (1989).

Maruyama, K.; Otagiri, M.; Shulman, S.G.: Binding characteristics of coumarin anticoagulants to human α_1-acid glycoprotein and human serum albumin. *Int. J. Pharmac.* **59**, 137-143 (1990).

Matzel, L.D.; Collin, C.; Alkon, D.L.: Biophysical and behavioral correlates of memory storage, degradation and reactivation. *Behav. Neurosci.* **106**(6), 954-963 (1992).

McNab, R.M., Aizawa, S.-I.: Bacterial motility and the flagellar motor. *Ann. Rev. Biophys. Bioeng.* **13**, 51-83 (1984).

McPhie. D.L.: Protein Kinase C as a nexus in memory formation. Ph.D. thesis, Georgetown University, Washington, D.C., USA (1994).

Metzinger, T. (ed.): *Conscious experience*. Schoeningh-Imprint Academic (1995).

Mezard, C.; Parisi, G.; Virasoro, P.: *Spin Glass Theory and Beyond*. World Scientific, Singapore (1987).

Miller, R.J.: Regulation of calcium homeostasis in neurons: the role of calcium-binding proteins. *Biochem. Soc. Trans.* **23**, 629-632 (1995).

Miura, K.; Kurosawa, Y.; Kanai, Y.: Calcium-binding activity of nucleobindin mediated by an EF-hand moyety. *Biochem. Biophys. Res. Comm.* **199**(3), 1388-1392 (1994).

Moncrief, N.D.; Kretsinger, R.H.; Goodman, M.: Evolution of EF-hand calcium-modulated proteins. I. Relationship based on amino acid sequence. *J. Mol. Evol.* **30**, 522-562 (1990).

Moore, G.D.; Ayabe, T.; Visconte, P.E.; Schultz, R.M.; Kopf, G.S.: Role of heterotrimeric and monomeric G-Proteins in sperm induced activation of mouse eggs. *Development* **120**, 3313-3323 (1994).

Moore, K.L.; Kinsey, W.H.: Effects of protein tyrosine kinase inhibitors on egg activation and fertilization-dependent protein tyrosine kinase activity. *Dev. Biol.* **168**, 1-10 (1995).

Moser, M.B.; Trammold, M.; Andersen, P.: An increase in dendritic spine density on hippocampus CA1 cells following spatial learning in adult rats suggests the formation of new synapses. *Proc. Natl. Acad. Sci. USA* **91**, 12673-12675 (1994).

Moshiach, S.; Nelson, T.J.; Sanchez-Andres, J.V.; Sakakibara, M.; Alkon, D.L.: *Brain Res.* **605**, 298-304 (1993).

Muller, W.; Connor, J.A.: Dendritic spines as individual neuronal compartments for synaptic Ca2+ responses. *Nature* **354**, 73-76 (1991).

Nagai, K.; Thoegersen, H.C.: Generation of β-globin by sequence specific proteolysis of a hybrid protein produced in Escherichia coli. *Nature* **309**, 810-812 (1984).

Nakayama, S.; Moncrief, N.D.; Kreitsinger, R.: Evolution of EF-hand calcium-modulated proteins. II. Domains of several subfamilies have diverse evolutionary histories. *J. Mol. Evol.* **34**, 416-448 (1992).

Neary, J.T.; Crow, T.; Alkon, D.L.: Change in a specific phosphoprotein band following associative learning in Hermissenda. *Nature* **293**, 658-661 (1981).

Nelson, T.J.; Alkon, D.L.: Prolonged RNA changes in the Hermissenda eye induced by classical conditioning. *Proc. Natl. Acad. Sci. USA* **85**, 7800-7804 (1988).

Nelson, T.J.; Alkon, D.L.: Specific high molecular weight mRNA induced by associative learning in Hermissenda. *Proc. Natl. Acad. Sci. USA* **87**, 269-273 (1990).

Nelson, T.J.; Collin, C.; Alkon, D.L.: Isolation of a G-protein that is modified by learning and reduces potassium channels in Hermissenda. *Science* **247**, 1479-1483 (1990)

Nelson, T.J.; Alkon, D.L.: GTP-binding proteins and potassium channels involved in synaptic plasticity and learning. *Mol. Neurobiol.* **5**, 315-328 (1991).

Nelson, T.J.; Sanchez-Andres, J.-V.; Schreus, B.G.; Alkon, D.L.: Classical conditioning-induced changes in low-molecular-weight GTP-binding proteins in rabbit hippocampus. *J. Neurochem.* **57**, 2065-2069 (1991).

Nelson, T.J.; Yoshioka, T.; Toyoshima, S.; Han, Y.-F.; Alkon, D.L.: Characterization of a GTP-binding protein implicated in both memory storage and interorganelle vescicle transport. *Proc. Natl. Acad. Sci. USA* **91**, 9287-9291 (1994).

Nelson, T.J.; Alkon, D.L.: Phosphorylation of the conditioning-associated GTP-binding protein cp20 by protein kinase C. *J. Neurochem.* **65**, 2350-2357 (1995).

Nelson, T.J.; Olds, J.L.; Kim, J.; Alkon, D.L.: Activation of DNA transcription in neuronal cells by an ARF-like GTP-binding protein. Submitted to *J. Biol. Chem.* (1996a).

Nelson, T.J; Cavallaro, S.; Yi, C.-L.; McPhie, D.; Schreus, B.G.; Gusev, P.A.; Favit, A.; Zohar, O.; Kim, J.-H.; Beushausen, S.; Ascoli, G.; Olds, J.L.; Neve, R.; Alkon, D.L.: Calexcitin, a signalling protein that binds calcium, inhinits potassium channels, and enhances membrane excitability. Submitted to *Proc. Natl. Acad. Sci. USA* (1996b).

Nishizuka, Y.: Intracellular signaling by hydrolysis of phospholipids and activation of protein kinase C. *Science* **258**, 607-614 (1992).

Nuwaysir, L.M.; Stultz, J.T.: Electrospray ionization mass spectrometry of phosphopeptides isolated by on-line immobilized metal-ion chromatography. *J. Am. Mass Spectr.* **4**, 662-669 (1993).

Ohlendiek, K., Lennarz, W.J.: Role of the sea urchin egg receptor for sperm in gamete interactions. *Trends Biochem. Sci.* **20**, 29-33 (1995).

Olds, J.L.; Anderson, M.L.; McPhie, D.L.; Staten, L.D.; Alkon, D.L.: Imaging of memory-specific changes in the distribution of protein kinase C in the hippocampus. *Science* **245**, 866-9 (1989).

Olds, J.L.; Golsky, S., McPhie, D.L.; Olton, D.; Mishkin, M; Alkon, D.L.: Discrimination learning alters the distribution of Protein Kinase C in the hyppocampus of rats. *J. Neurosci.* **10**, 3707-3713 (1990).

Olds, J.L.; Alkon, D.L.: A role for protein kinase C in associative learning. *The New Biol.* **3**, 27-35 (1991).

Olds, J.L.; Alkon, D.L.: Protein kinase C: a nexus in the biochemical events that undelie associative learning. *Acta Neurobiol. Exp.* **53**, 197-207 (1993).

Olds, J.L; Favit, A.; Ascoli, G.; Lester, D.S.; Rakow, T.; Alkon, D.L.: PKC in the sea urchin egg: molecular convergences and divergences of fertilization with associative memory. Abs. at the Neuroscience meeting, S. Diego USA (1995a).

Olds, J.L.; Favit, A.; Nelson, T.J.; Ascoli, G.; Gerstein, A.; Cameron, M.; Cameron, L.; De Barry, J.; Lester, D.S.; Rakow, T.; Yoshioka, T.; Freyberg, Z.; Baru, J.; Alkon, D.L.: Imaging protein kinase C activation in living sea urchin eggs after fertilization. *Dev. Biol.* **172**, 675-682 (1995b).

Olds, J.: Zip-coding the Dendritic Spine: A Biologically Plausible Solution to biological addressing within mammalian nerve cells. *Proc. Int. Neu. Net. Soc. Annu. Meeting.* **2**, 913-916 (1995)

Palkiewicz, P.; Zwiers, H.; Lorscheider, F.L.: ADP-ribosylation of brain neuronal proteins is altered by in vivo and in vitro exposure to inorganic mercury. *J. Neurochem.* **62**, 2049-2052 (1994).

Parker, F.S.: in *Application of Infrared, Raman and Resonance Raman Spectroscopy in Biochemistry*. Plenum Press, NY (1983).

Pavlov, I.P.: *Conditioned reflexes*. Anrep. Ed. Oxford Univ. Press, London (1910).

Peters, T.Jr.: Serum albumin. *Adv. Protein Chem.* **37**, 161-245 (1985).

Pottgiesser, J.; Maurer, P.; Mayer, U.; Nischt, R.; Mann, K.; Timpl, R.; Krieg, T.; Engel, J.: Changes in calcium and collagen IV binding caused by mutations in the EF hand and other domains of extracellular matrix protein BM-40 (SPARC, osteonectin). *J. Mol. Biol.* **238**, 563-574 (1994).

Rakow, T.L.; Shen, S.S.: Molecular cloning and characterization of protein kinase C from the sea-urchin Lytechinus Pictus. Devlop. *Growth and Differ.* **36**, 489-497 (1994).

Represa, A.; Deloulme, J.C.; Sensenbrenner, M; Ben-Ari, Y.; Baudier, J.: Neurogranin: immunocytochemical localization of a brain specific protein kinase C substrate. *J. Neurosci.* **10**, 3782-3791 (1990).

Robinson, P.J.; Liu, J.P.; Powell, K.A.; Fykse, E.M.; Sudhof, T.C.: Phosphorylation of dynamin I and synaptic-vesicle recycling. *Trends Neurosci.* **17**, 348-353 (1994).

Routtemberg, A.: A tale of two contingent protein kinase C activators: both neutral and aciic lipids regulate synaptic plasticity and informational storage. *Progr. Brain Res.* **89**, 249-261 (1991).

Sacktor, T.C.; Schwartz, J.H.: Sensitizing stimuli cause translocation of protein kinase C in Aplysia sensory neurons. *Proc. Natl. Acad. Sci. USA* **81**, 2036-2039 (1990).

Sagoo, J.K.; Chaplin, L.C.; Beeley, N.R.A.; Iaonnou, C.; Nesbitt, A.M.; Tendler, S.J.B.: Preparation and purification of 3-fluoro-tyrosine labelled tumor necrosis factor-α and preliminary ^{19}F n.m.r. investigation. *Int. J. Oncol.* **5**, 937-943 (1994).

Schneider, C.; Newman, R.A.; Sutherland, D.R.; Asser, U.; Greaves, M.F.: A one-step purification of membrane proteins using a high efficency immunomatrix. *J. Biol. Chem.* **257**, 10766-10769 (1982).

Schreus, B.G.: Classical conditioning of model systems: a behavioral review. *Psychobiol.* **17**(2), 145-156 (1989).

Schreus, B.G.; Sanchez-Andres, J.V.; Alkon, D.L.: Learning-specific differences in Purkinjie-cell dendrites of lobule HVI (Lobulus simplex): intracellular recording in a rabbit cerebellar slice. *Brain Res.* **548**, 18-22 (1991).

Segal, M.: Dendritic spines for neuroprotection: a hypothesis. *Trends Neurosci.* **18**, 468-471 (1995).

Segev, I.; Rinzel, J.; Shepherd, G.M.: *The Theoretical Foundation of Dendritic Function.* 115-122, MIT Press (1995).

Shao, X; Davletov, B.A.; Sutton, R.B.; Suedhof, T.C.; Rizo, J.: Bipartite Ca^{2+}-binding motif in C2 domains of synaptotagmin and protein kinase C. *Science* **273**, 248-251 (1996).

Shearman, M.S.; Sekiguchi , K.; Nishizuka, Y.: Modulation of ion channel activity: the key function of the protein kinase C enzyme family. *Pharm. Rev.* **41**, 211-237 (1989).

Shen, S.S.; Ricke, L.A.: Protein kinase C from sea urchin eggs. *Comp. Biochem. Physiol. [B]* **92**, 251-254 (1989).

Shen, S.S.; Buck, W.: A synthetic peptide of the pseudosubstrate domain of protein kinase C blocks cytoplasmic alkanization during activation of the sea urchin egg. *Devlop. Biol.* **140**, 272-280 (1990).

Sheterline, P. (ed.); Kawasaki, H.; Kretsinger, R.H.: Calcium binding proteins 1: EF-hands. *Protein Profile* **2**(4) 305-356 (1995).

Skene, J.H.P., Willard, M.: Axonally transported proteins associated with axon growth in rabbit central and peripheral nervous system. *J. Cell. Biol.* **89**, 96-103 (1981).

Smith, P.K.; Krohn, R.I.; Hermason, G.T.; Mallia, A.K.; Gartner, F.H.; Provezano, M.D.; Fujimoto, E.K.; Goeke, N.M.; Olson, B.J.; Klenk, D.C.: Measurement of protein using bincinchoninic acid. *Biochem.* **150**, 76-85 (1985).

Stabel, S.; Parker, P.J.: Protein Kinase C. *Pharmac. Ther.* **51**, 71-95 (1991).

Steinhardt, R.A.: Calcium regulation of the first cell cycle of the sea urchin embryo. *Ann. NY Acad. Sci.* **582**, 199-206 (1990a).

Steinhardt, R.A.: Intracellular free calcium and the first cell cycle of the sea-urchin embryo (Lytechinus pictus). *J. Reprod. Fertil.* Suppl. **42**, 191-197 (1990b).

Steward, O.; Falk, P.M.: Protein-synthetic machinery at postsynaptic sites during synaptogenesis: a quantitative study in the association between polyribosomes and developing synapses. *J. Neurosci.* **6**, 412-423 (1986).

Suelter, C.H.; DeLuca, M: How to prevent losses of protein by adsorption to glass and palastic. *Anal. Biochem.* **135**, 112-119 (1983).

Sunayashiki-Kusozaki, K.; Lester, D.S.; Schreus, B.G.; Alkon, D.L.: Associative learning potentiates protein kinase C activation in synaptosomes of the rabbit hyppocampus. *Proc. Natl. Acad. Sci. USA* **90**, 4286-4289 (1993).

Tanaka, K.; Tashiro, T.; Sekimoto, S.; Komiya, Y.: Axonal transport of actin and actin-binding proteins in the rat sciatic nerve. *Neurosci. Res.* **19**, 295-302 (1994).

Travè, G.; Pastore, A.; Hyvoenen, M; Saraste, M: The C-terminal domain of α-spectrin is structurally related to calmodulin. *Eur. J. Biochem.* **227**, 35-42 (1995).

Van Hooff, C.O.M.; De Graan, P.N.E.; Oestreicher, A.B.; Gispen, W.H.: Muscarinic receptor activation stimulates B50/GAP-43 phosphorylation in isolated nerve growth cones. *J. Neurosci.* **8**, 1789-1795 (1989).

Viegi, A.: Dicroismo circolare nella caratterizzazione della siero albumina umana nativa e chimicamente modificata. Ph.D. thesis, University of Pisa (1994).

Yloenen, J.; Virtanen, T.; Horsmanheimo, L.; Parkkinen, S.; Pelkonen, J.; Maentyjaervi, R.: Affinity purification of the major bovine allergen by a novel monoclonal antibody. *J. Allergy Clin. Immunol.* **93**, 851-858 (1994).

Walker, J.E.: Lysine residue 199 of human serum albumin is modified by acetyl-salicylic acid. *FEBS Letters* **66**, 173-175 (1976).

Wang, C.-K.; Mani, R.S.; Kay, C.M.; Cheung, H.C.: Conformation and dynamics of bovine brain S-100a protein determined by fluorescence spectroscopy. *Biochem.* **31**, 4289-4295 (1992).

Wang, L.; Ross, J.: Physical modeling of neural networks. In *Methods in Neuroscience: Computers and Computations in the Neurosciences*. P.M. Conn ed., Academic Press San Diego (1992).

Warren, T.G.: Use of affinity-purified antibodies for protein isolation. *Peptides* **13**, 74-78 (1994).

Warren, T.G.; Hippenmeyer, P.J.; Meyer, D.M.; Reitz, D.M; Rowold, E.Jr.; Carron, C.P.: High-level expression of biologically active, soluble forms of ICAM-1 in a novel mammalian-cell expression system. *Prot. Expr. Purif.* **5**, 498-508 (1994).

Watanabe, T.; Ono, Y.; Taniyama, Y.; Hazama, K.; Igarashi, K.; Ogita, K.; Kikkawa, U.; Nishizuka, Y.; Cell division arrest induced by phorbol ester in CHO cells overexpressing protein kinase C-delta subspecies. *Proc. Natl. Acad. Sci. USA* **89**, 10159-10163 (1992).

Weber, C.; Lee, V.D.; Chazin, W.J.; Huang, B.: High level expression in Escherichia coli and characterization of the EF-hand calcium-binding protein caltractin. *J. Biol. Chem.* **269**, 15795-15802 (1994).

Woody, R.W.: Circular dichroism of peptides. *The Peptides* **7**, 15-114 (1985).

Zhao, Y.; Pokutta, S.; Maurer, P.; Lindt, M.; Franklin, R.M.; Kappes, B.: Calcium-binding properties of a calcium-dependent protein kinase from plasmodium falciparum

and the significance of individual calcium-binding sites for kinase activation. *Biochem.* **33**, 3714-3721 (1994).

Zhou, W.; Takuwa, N.; Kumada, M.; Takuwa, Y.: Protein kinase C-mediated bidirectional regulation of DNA synthesis, RB protein phosphorylation, and cyclin-dependent kinases in human vascular endothelial cells. *J. Biol. Chem.* **268**, 23041-23048 (1993).

Zidovetzki, R.; Lester, D.S.: The mechanism of activation of protein kinase C: a biophysical perspective. *Biochim. Biophys. Acta* **1134**, 261-272 (1992).

Zimmer, D.B.; Cornwall, E.H.; Landar, A.; Song, W.: The S100 protein family: history, function, and expression. *Brain Res. Bull.* **37**(4), 417-429 (1995).

Zwiers, H.; Schotman, P. and Gispen, W.H.: Purification and some characteristics of an ACTH-sensitive protein kinase and its substrate protein in rat brain. *J. Neurochem.* **34** (6), 1688-1699 (1980).

Elenco delle Tesi di perfezionamento della Classe di Scienze
pubblicate dall'Anno Accademico 1992/93

HISAO FUJITA YASHIMA, *Equations de Navier-Stokes stochastiques non homogènes et applications*, 1992.

GIORGIO GAMBERINI, *The minimal supersymmetric standard model and its phenomenological implications*, 1993.

CHIARA DE FABRITIIS, *Actions of Holomorphic Maps on Spaces of Holomorphic Functions*, 1994.

CARLO PETRONIO, *Standard Spines and 3-Manifolds*, 1995.

MARCO MANETTI, *Degenerations of Algebraic Surfaces and Applications to Moduli Problems*, 1995.

ILARIA DAMIANI, *Untwisted Affine Quantum Algebras: the Highest Coefficient of* det H_η *and the Center at Odd Roots of 1*, 1995.

FABRIZIO CEI, *Search for Neutrinos from Stellar Gravitational Collapse with the MACRO Experiment at Gran Sasso*, 1995.

ALEXANDRE SHLAPUNOV, *Green's Integrals and Their Applications to Elliptic Systems*, 1996.

ROBERTO TAURASO, *Periodic Points for Expanding Maps and for Their Extensions*, 1996.

YURI BOZZI, *A study on the activity-dependent expression of neurotrophic factors in the rat visual system*, 1997.

MARIA LUISA CHIOFALO, *Screening effects in bipolaron theory and high-temperature superconductivity*, 1997.

DOMENICO M. CARLUCCI, *On Spin Glass Theory Beyond Mean Field*, 1998.

RENATA SCOGNAMILLO, *Principal G-bundles and abelian varieties: the Hitchin system*, 1998.

GIACOMO LENZI, *The MU-calculus and the Hierarchy Problem*, 1998.

GIORGIO ASCOLI, *Biochemical and spectroscopic characterization of CP20, a protein involved in synaptic plasticity mechanism*, 1998.

"CompoMat" Loc. Braccone, 02040 Configni (RI), Italy
Finito di stampare nel maggio 1998